T0259007

COMPUTATIONAL INTELLIGENCE APPLICATIONS FOR SOFTWARE ENGINEERING PROBLEMS

COMPUTATIONAL INTELLIGENCE APPLICATIONS FOR SOFTWARE ENGINEERING PROBLEMS

Edited by
Parma Nand, PhD
Nitin Rakesh, PhD
Arun Prakash Agrawal, PhD
Vishal Jain, PhD

APPLE
ACADEMIC
PRESS

First edition published 2023

Apple Academic Press Inc.
1265 Goldenrod Circle, NE,
Palm Bay, FL 32905 USA

760 Laurentian Drive, Unit 19,
Burlington, ON L7N 0A4, CANADA

CRC Press
6000 Broken Sound Parkway NW,
Suite 300, Boca Raton, FL 33487-2742 USA

4 Park Square, Milton Park,
Abingdon, Oxon, OX14 4RN UK

© 2023 by Apple Academic Press, Inc.

Apple Academic Press exclusively co-publishes with CRC Press, an imprint of Taylor & Francis Group, LLC

Library and Archives Canada Cataloguing in Publication

Title: Computational intelligence applications for software engineering problems / edited by Parma Nand, PhD, Rakesh Nitin, PhD, Arun Prakash Agrawal, PhD, Vishal Jain, PhD.

Names: Nand, Parma (Computer scientist), editor. | Nitin, Rakesh, editor. | Agrawal, Arun Prakash, editor. | Jain, Vishal, 1983- editor.

Description: First edition. | Includes bibliographical references and index.

Identifiers: Canadiana (print) 20220267731 | Canadiana (ebook) 20220267855 | ISBN 9781774910467 (hardcover) | ISBN 9781774910474 (softcover) | ISBN 9781003283195 (ebook)

Subjects: LCSH: Software engineering. | LCSH: Computational intelligence.

Classification: LCC QA76.758 .C66 2023 | DDC 005.1—dc23

Library of Congress Cataloging-in-Publication Data

Names: Nand, Parma (Computer scientist), editor. | Nitin, Rakesh, editor. | Agrawal, Arun Prakash, editor. | Jain, Vishal, 1983- editor.

Title: Computational intelligence applications for software engineering problems / edited by Parma Nand, PhD, Rakesh Nitin, PhD, Arun Prakash Agrawal, PhD, Vishal Jain, PhD.

Description: First edition. | Palm Bay, FL : Apple Academic Press, Inc.; Boca Raton, FL : CRC Press, 2023. | Includes bibliographical references and index. | Summary: "This new volume explores the computational intelligence techniques necessary to carry out different software engineering tasks. Software undergoes various stages before deployment, such as requirements elicitation, software designing, software project planning, software coding, and software testing and maintenance. Every stage is bundled with a number of tasks or activities to be performed. Due to the large and complex nature of software, these tasks become more costly and error prone. This volume aims to help meet these challenges by presenting new research and practical applications in intelligent techniques in the field of software engineering. Computational Intelligence Applications for Software Engineering Problems discusses techniques and presents case studies to solve engineering challenges using machine learning, deep learning, fuzzy-logic-based computation, statistical modeling, invasive weed meta-heuristic algorithms, artificial intelligence, the DevOps model, time series forecasting models, and more. This volume will be helpful to software engineers, researchers, and faculty and advanced students working on intelligent techniques in the field of software engineering"-- Provided by publisher.

Identifiers: LCCN 2022026648 (print) | LCCN 2022026649 (ebook) | ISBN 9781774910467 (hbk) | ISBN 9781774910474 (pbk) | ISBN 9781003283195 (ebk)

Subjects: LCSH: Software engineering--Data processing. | Artificial intelligence.

Classification: LCC QA76.758 .C628 2023 (print) | LCC QA76.758 (ebook) | DDC 005.1--dc23/eng/20220805

LC record available at https://lccn.loc.gov/2022026648

LC ebook record available at https://lccn.loc.gov/2022026649

ISBN: 978-1-77491-046-7 (hbk)
ISBN: 978-1-77491-047-4 (pbk)
ISBN: 978-1-00328-319-5 (ebk)

About the Editors

Parma Nand, PhD

Parma Nand, PhD, has more than 27 years of experience both in industry and academia. He has received various awards, including a best teacher award from the Government of India, best student's project guide award from Microsoft in 2015, and best faculty award from Cognizant in 2016. He has successfully completed government-funded projects and has spearheaded last five IEEE International Conferences on Computing, Communication and Automation (ICCCA), IEEE Student's Chapters, Technovation Hackathon 2019, Technovation Hackathon 2020, International Conference on Computing, Communication, and Intelligent Systems (ICCCIS-2021). He is a member of the Executive Council of IEEE, Uttar Pradesh section (R-10), member of the Executive Committee of IEEE Computer and Signal Processing Society, member Executive India Council of the Computer Society, and member of the Executive Council Computer Society of India, Noida section, and has acted as an observer at many IEEE conferences. He also has active memberships in other professional organizations, including ACM, IEEE, CSI, ACEEE, ISOC, IAENG, and IASCIT. He is a lifetime member of the Soft Computing Research Society (SCRS) and ISTE. Dr. Nand has delivered many invited and keynote talks at international and national conferences, workshops, and seminars in India and abroad. He has published more than 150 papers in peer-reviewed international and national journals and conferences. He has also published number of book chapters in reputed publications from Springer and others. He has reviewed a number of books for publishers including Tata McGraw-Hill and Galgotias Publications and has reviewed papers for international journals. He is an active member of advisory/technical program committees of reputed international and national conferences. His research interests include computer graphics, algorithms, distributed computing, and wireless and sensor networking. Dr. Nand earned his PhD in Computer Science and Engineering from IIT Roorkee and his MTech and BTech in Computer Science and Engineering from IIT Delhi, India.

Nitin Rakesh, PhD

Nitin Rakesh, PhD, is an experienced professional in the field of computer science. Currently, Dr. Nitin is Head of the Computer Science & Engineering Department for BTech/MTech (CSE/IT), BTech CSE-IBM Specializations, BTech CSE-I Nurture, BCA/MCA, BSc/MSc-CS at, Sharda University, Greater Noida, U.P., India. He is also working with various other departments for enhancement, research and academic initiatives, industrial interfacing, and other major developments. He is an active member of professional societies that include IEEE (USA), ACM, and SIAM (USA) and is a life member of CSI and other professional societies. He is reviewer for several prestigious journals, including *IEEE Transactions on Vehicular Technology, The Computer Journal,* and other Scopus-indexed international journals. He has over 100 publications in Scopus-indexed/SCI/high-impact journals and international conferences. His research works is on network coding, interconnection networks and architectures, and online phantom transactions. Dr. Nitin has been awarded several awards for best published paper, highest cited author, best guided student thesis, and many others. Dr. Nitin has been instrumental in various industrial interfacing for academic and research at his previous assignments at various organizations (Amity University, Jaypee University, Galgotias, and others). Dr. Nitin is guiding eight PhD students at various universities and institutes and has successfully guided several MTech and BTech students. Dr. Nitin was contributor to various prestigious accreditations, including NAAC, NBA, QAA, WASC, UGC, IAU, IET, and others. He has initiated an IoT and network lab at Sharda University, which will be a new technology initiative for graduate, postgraduate, and doctoral students.

He has a PhD in computer science and engineering with network coding as his specialization. Dr. Nitin has Master of Technology degree in computer science and engineering and received a Bachelor of Technology in information technology. Dr. Nitin is a recipient of IBM Drona Award and is Top 10 State award winner.

Arun Prakash Agrawal, PhD

Prof. Arun Prakash Agrawal is Professor with the Department of Computer Science and Engineering at Sharda University, Greater Noida, India. Prior to his current assignment, he has served many academic institutions, including Amity University, Noida. He has several research papers to his credit in refereed journals and conferences of international repute in India and abroad, including *Journal of Neural Computing and Applications,* published by Springer. He has also taught short-term courses at Swinburne University of Technology, Melbourne, Australia, and Amity Global Business School, Singapore. His research interests include machine learning, software testing, artificial intelligence, and soft computing. He has obtained his PhD and master's in Computer Science and Engineering from Guru Gobind Singh Indraprastha University, New Delhi, India. He was the gold medalist of his batch.

Vishal Jain, PhD

Vishal Jain, PhD, is Associate Professor in the Department of Computer Science and Engineering, School of Engineering and Technology, Sharda University, Greater Noida, U.P., India. Before that, he has worked for several years as an associate professor at Bharati Vidyapeeth's Institute of Computer Applications and Management (BVICAM), New Delhi. He has more than 14 years of experience in the academics. He has more than 400 research citation indices with Google Scholar (h-index score 10 and i-10 index 11). He has authored more than 85 research papers in reputed conferences and journals, including the Web of Science and Scopus. He has authored and edited more than 10 books with various reputed publishers, including Springer, Apple Academic Press, CRC, Taylor and Francis Group, Scrivener, Wiley, Emerald, and IGI-Global. His research areas include information retrieval, semantic web, ontology engineering, data mining, ad hoc networks, and sensor networks. He received a Young Active Member Award for the year 2012–2013 from the Computer Society of India and Best Faculty Award for the year 2017 and Best Researcher Award for the year 2019 from BVICAM, New Delhi. He holds PhD (CSE), MTech (CSE), MBA (HR), MCA, MCP, and CCNA.

Contents

Contributors

Alankrita Aggarwal
Department of Computer Science and Engineering, IKGPTU, Kapurthala 144603, Punjab, India

Vikramaditya Dave
Department of Electrical Engineering, College of Technology and Engineering, Udaipur 313001, Rajasthan, India

Kanwalvir Singh Dhindsa
Department of Computer Science and Engineering, Baba Banda Singh Bahadur Engineering College, Fatehgarh Sahib 140406, Punjab, India

Jayesh M. Dhodiya
Department of Applied Mathematics and Humanities, S. V. National Institute of Technology, Surat, India

Shruti K. Dixit
Electronics and Communication Department, Rajiv Gandhi, Proudyogiki Vishwavidyalaya (RGPV), Sagar Institute of Research and Technology, Bhopal, Madhya Pradesh, India

Spandana Gowda
Department of Electrical and Electronics Engineering, Jain (Deemed-to-be University), Bangalore 560041, Karnataka, India

Somya Goyal
Department of Computer and Communication Engineering, Manipal University Jaipur, Jaipur, India
Department of Computer Science and Engineering, Guru Jambheshwar University of Science & Technology, Hisar, India; E-mail: somyagoyal1988@gmail.com

Ishwarappa Kalbandi
Computer Engineering, Dr. D.Y. Patil Institute of Engineering, Management & Research Akurdi, Pune, India; E-mail: ishwar.kalbandi@gmail.com

Navneet Kaur
Electronics and Communication Department, Rajiv Gandhi, Proudyogiki Vishwavidyalaya (RGPV), Sagar Institute of Research and Technology, Bhopal, Madhya Pradesh, India

Sujit Kumar
Department of Electrical and Electronics Engineering, Jain (Deemed-to-be University), Bangalore 560041, Karnataka, India; E-mail: sujitvj.kumar@gmail.com

D. K. Mishra
Department of Engineering Mathematics & Computing, Madhav Institute of Technology & Science, Gwalior, India

Mohana
Electronics and Telecommunication Engineering, RV College of Engineering, Bangalore 560059, India

Ambika N.
Department of Computer science and Applications, St. Francis College, Bangalore, India;
E-mail: ambika.nagaraj76@gmail.com

Naresh Kumar Nagwani
Department of Computer Science and Engineering, National Institute of Technology, Raipur,
Chhattisgarh, India

V. Lakshman Narayana
Department of IT, Vignan's Nirula Institute of Technology & Science for Women,
Andhra Pradesh, India

Rama Ranjan Panda
Department of Computer Science and Engineering, National Institute of Technology, Raipur,
Chhattisgarh, India

R. S. M. Lakshmi Patibandla
Department of IT, Vignan's Foundation for Science, Technology, and Research, Andhra Pradesh,
India

B. Tarakeswara Rao
Department of CSE, Kallam Haranadhareddy Institute of Technology, Guntur, Andhra Pradesh, India

Shalini Sahay
Electronics and Communication Department, Rajiv Gandhi, Proudyogiki Vishwavidyalaya (RGPV),
Sagar Institute of Research and Technology, Bhopal, Madhya Pradesh, India

Divya Sharma
G. D. Goenka University, Gurugram, India; E-mail: divya.07sharma@gmail.com

Ganga Sharma
G. D. Goenka University, Gurugram, India

Vikas Shinde
Department of Engineering Mathematics & Computing, Madhav Institute of Technology & Science,
Gwalior, India

Shweta Shrivastava
Department of Engineering Mathematics & Computing, Madhav Institute of Technology & Science,
Gwalior, India

V. Sesha Srinivas
Department of Information Technology, RVR & JC College of Engineering, Andhra Pradesh, India

P. K. Suri
Department of Computer Science & Applications, Kurukshetra University, Kurukshetra 136038,
Haryana, India

Anita R. Tailor
Department of Mathematics, Navyug Science College, Surat, India;
E-mail: anitatailor_185@outlook.com

Vidhi Vig
Department of Computer Science, S.G.T.B. Khalsa College, University of Delhi, North Campus,
New Delhi 110007, India; E-mail: vidhi.ipu@gmail.com

Abbreviations

AAA	authentication, authorization and accounting
ABC	artificial bee colony
ABE	analogy-based estimation
ACO	ant colony optimization
AI	artificial intelligence
AIC	Akaike information criterion
AIS	artificial immune system
ALs	aspiration levels
ANN	artificial neural network
APN	access point names
ARIMA	auto-regressive integrated moving average
ASP	application software package
BBO	biogeography-based optimization
CCA	credit control answer
CCR	credit control request
CI	computational intelligence
CLI	command-line interface
CMO	crisp multi-objective
CMP	cloud management platform
CNN	convolutional neural networks
COTS	commercial-off-the-shelf
CQI	channel quality indicator
DA	defect acceptance
DDE	dynamic diameter engine
DE	differential evolution
DevOps	development and operations
DIWO	discrete invasive weed optimization
DL	deep learning
DM	decision maker
DR	defect rejection
DRL	deep reinforcement learning
DSC	diameter signaling controller
DTLS	datagram transport layer security

EASs	edge access systems
EC	evolutionary computation
ECRSs	emergency call routing services
ECUR	event charging with unit reservation
EMF	exponential membership function
EPS	enhanced presence service
ERP	enterprise resource planning
ESs	efficient solutions
ETSI	European Telecommunication Standard Institute
EV	effort variance
FF	feed forward
FLC	fuzzy logic controller
GA	genetic algorithm
GP	genetic programming
GPGPU	general-purpose-graphics processing unit
GSM	global system for mobile
HS	harmonic search
HSS	home subscriber server
IEC	Immediate Event Charging
IFS	intuitionistic fuzzy sets
IMS	IP multimedia subsystem
ILP	integer linear programming
IoT	internet of things
IR	intricacy rank
IST	information and software technology
IWO	invasive weed optimization
KPI	key performance indicator
LN	large negative
LP	large positive
LP	linear programming
LTE	long-term evolution
MAD	mean absolute deviation
MAE	mean absolute error
MAPE	mean absolute percent error
MCC	Mathew's correlation coefficient
ML	machine learning
MME	mobility management entity
MN	medium negative

MOOM	multi-objective optimization model
MP	medium positive
MSE	mean square error
MSS	modular software system
MSR	mining software repositories
NAS	network access server
NAS	nonaccess stratum
NE	network element
NIS	negative ideal solution
NYB	New York City births
OCSs	online charging systems
OEE	overall equipment effectiveness
OfCS	off-line charging system
OR	operational research
OS	operating system
OSI	open systems interconnection
PCRF	policy and charging control entity/function
PCRF	policy and charging rules function
PdM	predictive maintenance
PDN	packet data network
PDN-GW	packet data network-gateway
PGW	PDN Gateway
PF	polymorphism factor
PIS	positive ideal solution
PSO	particle swarm optimization
QoS	quality of service
RADIUS	remote authentication dial in user service
RAN	radio access network
RBN	radial basis network
RIDE	ROBOT integrated development environment
RMSE	root mean square error
RNN	recurrent neural networks
SA	simulated annealing
SBO	search-based optimization
SBSE	search-based software engineering
SC	scope change
SCP	secure copy protocol
SCTP	stream control transmission protocol

SCUR	charging with unit reservation
SD	software developers
SDLC	software development life cycle
SDP	software defect prediction
SE	software engineering
SEE	software effort estimation
SFTP	secure file transfer protocol
SGW	serving gateway
SI	swarm intelligence
SM	software measurement
SM GUI	service manager graphic user interface
SONOP	single objective nonlinear optimization problem
SPs	shape parameters
SSH	secure shell
SV	schedule variance
SVM	support vector machine
TCP	test case productivity
TCP/IP	transmission control protocol/internet protocol
TEP	test execution productivity
TLS	transport layer security
TPM	total productive maintenance
TS	tabu search
UDP	user datagram protocol
UE	user equipment
UMO	uncertain multi-objective
UMTS	universal mobile telecommunications system
USIM	universal subscriber identity module
VNF	virtual network functions

Preface

Size and complexity of software systems are increasing day by day, and so are the challenges associated with them. Before becoming obsolete, any software undergoes a number of stages during its lifecycle, such as requirements engineering, design, coding, testing, and maintenance to name a few. Each of these stages accommodates a number of costly and error-prone activities that are performed. Computational intelligence techniques can be applied to carry out these activities effectively and efficiently. Computational intelligence techniques are aimed at providing better and optimal solutions to real-world complex optimization problems in reasonable time limit. These are also closely related to artificial intelligence (AI) and incorporate heuristic as well as metaheuristic algorithms. These approaches have successfully been applied to solve real-world problems in various application domains such as healthcare, bioinformatics, civil engineering, computer networks, scheduling, software project planning, resource allocation, and forecasting, to name a few. These techniques have also attracted researchers in the software engineering domain and have been successful in prioritization of requirements, size and cost estimation of software to be developed, software defect prediction and reliability assessment, test case prioritization and vulnerability prediction, and many more.

Size of solution space in such domains is very large, and intelligent behavior shown by computational intelligence techniques including evolutionary algorithms, machine learning algorithms, and metaheuristic algorithms find appropriate application of these approaches.

Machine learning approaches are, however, constrained by the availability of huge amount of data to extract knowledge and to build and train the model. A metaheuristic algorithm, on the other hand, is a high-level, iterative process. Exploration and exploitation are the two key characteristics of any metaheuristic algorithm that guides and manipulates an underlying heuristic algorithm to find an optimal solution to the problem at hand. The underlying heuristic algorithm can be either a local search, greedy approach, or some other low- or high-level procedure. Metaheuristic algorithms have been able to find near-optimal solutions to a large variety of problems with high accuracy and limited resources in a reasonable amount of time.

Having found the importance of such techniques, this book invited researchers, academicians, and professionals to contribute chapters expressing their ideas and research toward the application of computational intelligence techniques in the field of software engineering.

Chapter 1 discusses a proposed technique based on code analytics put together from reenactment designs and simulation patterns with the investigation of deep code evaluation with the analytics of the respective performance using statistical simulation and which is capable of achieving better performance for predictive analytics after multiple simulation attempts.

Chapter 2 gives detailed insights into each challenge by giving suitable case studies from real-life experiments. With each case study, a proposed solution is also offered.

Chapter 3 describes the problem that can be used to direct upcoming studies in software engineering and AI societies. Together, we will encourage an incredible amount of corporations to start capturing the benefit of the high latent of AI tools.

Chapter 4 focuses a genetic algorithm (GA)-based approach to find the solution of uncertain multi-objective commercial-off-the-shelf (COTS) product selection problem with triangular and/or trapezoidal numbers in objective functions subject to several pragmatic constraints.

Chapter 5 discusses the problem faced by software industries and how these problems can be solved using fuzzy logic techniques. This chapter enlightens on the different areas of software bugs and their improvement using fuzzy logic based techniques.

Chapter 6 provides practical orientation to readers on the implementation of deep learning technique to software measurements.

Chapter 7 presents an extensive summary of the state of art of time series forecasting models and summarizes the evolution of models in the arena since early 2000s.

Chapter 8 illustrates AI's role in predictive maintenance, the need for AI application in manufacturing, a proposed model for predictive maintenance of Industry4.0, and finally a discussion on accurate results.

Chapter 9 evaluates the efficacy of IWO via a collection of multi-model benchmark functions.

Chapter 10 explores the use of computational intelligence techniques (specifically fuzzy logic, artificial neural network, and particle swarm optimization) for classification of data.

Chapter 11 provides an enhanced study of intelligence architecture.

Chapter 12 gives a systematic literature review of search-based software engineering techniques for code modularization/remodularization.

Chapter 13 gives an overview of automation of framework using DevOps model to deliver DDE software.

A Statistical Experimentation Approach for Software Quality Management and Defect Evaluations

ALANKRITA AGGARWAL[1*], KANWALVIR SINGH DHINDSA[2], and P. K. SURI[3]

[1]*Department of Computer Science and Engineering, IKGPTU, Kapurthala 144603, Punjab , India*

[2]*Department of Computer Science and Engineering, Baba Banda Singh Bahadur Engineering College, Fatehgarh Sahib 140406, Punjab, India*

[3]*Department of Computer Science & Applications, Kurukshetra University, Kurukshetra 136038 Haryana, India*

*Corresponding author. E-mail: alankrita.agg@gmail.com

ABSTRACT

Software development and metrics areas metrics lack information concerning metrics on what is applicable and what is actually practiced. In this work, the goal is to research software metrics about the source code to direct software quality measurement. Assessment of formulations is prearranged from two eras: first, when focus was on metrics based on complexity of the code and, second, when the center was on metrics based on object-oriented systems. The proposed technique for code analytics based on re-enactment designs and simulation patterns investigating deep code evaluation with the analytics of the respective performance. These

Computational Intelligence Applications for Software Engineering Problems. Parma Nand, PhD, Rakesh Nitin, PhD, Arun Prakash Agrawal, PhD & Vishal Jain, PhD (Eds.)

performances used statistical simulation and are capable doing better for predictive analytics after multiple simulation attempts. To improve the software quality by one-fourth of the time taken, work is done to minimize the time taken to achieve accuracy compared with classical methods.

1.1   FOREWORD AND INTRODUCTION

Quality is an experience in which number of variables is involved depending on individual behavior, and it is not an easy task to control it. Metric or formulation methods are used to gauge and enumerate such variables. Descriptions of software metric or formulations can be used in research works.[1] Daskalantonakis found that metrics and measures are the best motivators to quantify; for example, to find a mathematical value for product and process attributes. In return, these values can be compared with standards applicable in the organization to draw conclusions about the quality of metrics used.[2] Till now, there is no formal procedure to measure the quality of the source code. According to Spinellis[3], source code quality is defined by two methods: technically and conceptually. Technological description associated with source code styles stresses on readability and language-explicit convention for maintenance of the software source code that includes updating and debugging of the code. As a result, conceptual explanation is associated with the logical structure of code to be controllable and quality features like readability, maintainability, testing, portability, and complexity.[4] This work will evaluate up to date software metrics applied on the source code to analyze the existing software metrics but it does not cover performance metrics and productivity. Also, verification of the development of metrics and source code quality could not survive in the process.[5]

In the latest research work, large amount of actions have been there for capturing software attributes quantitatively. Although some of them stay alive and are practically used in the industry, many troubles are accountable for metric failures that are identified.[6] If we try to pick out the problems for the analysis of metric or formulations, then it can be categorized as:

- Metric or formulations automation.
- Metric or formulations validation.

After the passage of time, production tools for metric or formula-tions drawing out are recognized because a large number of metric or formulations that do not have a clear definition are developed. In general, for a set of programmer languages, a code metric or formulation is made in larger context and is validated[7], and problems about validation are analyzed.

1.2 BRIEF SUREVY ON METRICS

This section presents a survey on the software metric or formulations. The program volume represented as V gives contents of the program that are calculated mathematically as program length multiplied by 2-base logarithm of vocabulary size (n) and is given by the following formula:

$$V = N \times \log 2(n) \tag{1.1}$$

Halstead's volume represented as HV gives length of lines of code used in implementing of the algorithm. The calculation of HV is dependent on operations and operands in algorithm and more responsive to measurement of LOC than complete source code.[8] The volume of a function must be at least 20 and maximum 1000. The parameterless one-line function is not empty but has a volume of about 20, and a function greater than 1000 has a larger volume tells that function to do many things.

The volume of file V(G) is between 100 and 8000 based on the number of files whose LOC are within suggested limits and can be used for cross-checking.

The intricacy rank of the program compares all exclusive operators in the program. D is the ratio between all operands and all unique operands. It is advisable not to use the same operands many times in the program to avoid errors.[9]

$$D = (n1 \div 2) \times (N2 \div n2) \tag{1.2}$$

Program level (L) is defined as converse of fault proneness of program, that is, low-level programs are additional prone to errors than the high-level program.

$$L = 1/D \tag{1.3}$$

Daskalantonakis adopted definition and also gave a number of definitions when research work is carried out connected to software metric or formulations and here software metric or formulations is to enumerate features in software projects, processes, and products.

A numerical value for measuring the software measurement of the software product as well as process attributes and comparing with each other in an organization to draw a conclusion about the quality of the software process[10].

Somerville also classifies metric or formulations in two classes:

(a) Control metric or formulations connected with software process.
(b) Predict metric or formulations connected with the software product.

Here the focus is both on predict metric or on formulations as they forecast the metric or formulations and measure software on the basis of static and dynamic features. For example, compute complication of code, instability, coupling/cohesion, inheritance among product attributes and examine quality, and suggest improvement for improving reusability, effort, and testability.

The secondary metric or formulations provides the details how the front end and database system is working. For example, fan-in/fan-out feature is to the calculate an estimate of worst-case complexity of the section and what is the strength of component's communication relationship among them.[11]

If we talk about the validity of formulation, it is ignored by the system design and there is a high association among values and counting of error when the system consists of less number of modules are analyzed.

One more complexity formulation given by McClure focuses on complexity of control structures and control variables to invocation methods. Invocation complexity or higher invocation complexity is assigned to component that invokes unconditionally on condition that the variables are customized by associates or descendents, respectively.[12]

Woodfield also mentioned a complexity system metric where each module context enquired by component or affects by another component and reviewed and from previous reviews less effort is taken by earlier ones. A review is decreasing the function weight complexity, and summation of all weights is the measure assigned to module and metric applied for multiprocedure programs and reports about tools implementing those are found.[13]

Henry and Kafura in 1981 produced a detailed system complexity metric that determines the density of complexity depending on two factors:

(a) Density of the procedure code.
(b) Density of the procedure's connections to surroundings.

Yin and Winchester formed a formulation that monitors all information flow automated instead of just flow across level boundaries.[14]

More definitions, like flows definition and modules definition, given by the researchers are more puzzling. Problems in Henry and Kafura's approach are of validation as algebraic expression is arbitrary, and application of parametric tests is questionable by using formulation.[15]

The CK metric or formulations is most referenced, and commercial is a collection of tools available at the time of writing and used by many researchers in industry and academies.

Lorenz and Kidd developed metric or formulations on object-oriented systems and divided classes based in four categories:

- Size: metric that is size oriented for OO classes focus on counting attributes and operations.
- Inheritance: metrics based on inheritance focus on which operations are reused in the hierarchy class.
- Internals: internal classes look at cohesion and issues of leaning of code.
- Externals: external metric inspect reusability and coupling.

Sometimes CK metric or formulations is more famous Lorenz and Kidd's metric or formulations as they include OO attributes in their analysis. Till today, no such publication has been published on the metrics. Codes written in Smalltalk and C++ were tested on a tool called OO formulation was developed.

To measure inheritance metric like information hiding, polymorphism related to software quality, and development productivity and validation is dubious for polymorphism factor (PF) formulation. If PF is not the valid system, it is without inheritance, discontinuous, and not defined. MOODKIT is a tool for metric drawing out from the source code that supports collection for C++/ Smalltalk/Eiffel programs codes.

Empirical validation's aim is to build quality models of OO designs then coupling is an important structural dimension. Therefore, the metric or formulations is measurement of coupling between classes and a tool for extraction of this metric or formulations can be built or found. The research is going on since 1990s, and OO metric or formulations was created by analyzing work, validating metric, or formulations were published. After the progress of 2000, there are few reports on software metric or formulations. The proposed metric or formulation is not based in the classical metric or formulation frameworks.[16] The work of Chatzigeorgiou is pioneering as applying web algorithmic verifies relation between design classes and not using existents metric or formulations. Chatzigeorgiou validates metric or formulation comparing to OO metric. Analysis for verification of linked classes and capability to judge both incoming and outgoing flows of messages done by L&K metric and can draw out find a tool for the metric or formulations.

Nowadays, most of the business as well as social media services are available online on assorted virtual platforms including mobile apps and web portals. These online platforms provide a higher degree of accuracy and performance services without delay. Even small business ventures and entrepreneurs are now developing and launching their mobile apps so that their visibility and accessibility can be increased with the 24 × 7 availability as well as communication with the customers. With the development and launching of mobile apps, there exist so many risk factors and vulnerabilities with each app because it is always required to test the mobile application on assorted parameters. Simply deployment of a mobile app without proper and rigorous testing can be dangerous for the vendor.[17] Many a times it happens that the apps of e-wallets are exploited and the app users lose their money. Such situations hurt the overall reputation and reliability factors of the seller or the company owning that app. It is always desired that the mobile app should be tested with assorted types of attacks so that the prior information on the behavior of app can be obtained. Once the app is tested on assorted types of traffic, it becomes very useful to evaluate the performance of app without actual launching the app on cyberspace.

Researchers suggested the number of predictions techniques like neural networks, classification, etc.[18] Clusters can be made using the cloud spark of various predictions patters received after different simulations.[19–21] An improved software improvement model can be designed using different metrics and methods.[22,23] Moreover refactoring of object-oriented methods can be used by an effective team.[24] A different scheme by using nature-inspired behavior-driven methods defects can be minimized proposed.[25] For improving performance random forest algorithms are used and voting criteria can be implemented for defect minimization.[26] Different models and frameworks using a biological approach and computing methods are studied for creation of an effective futuristic model.[27–30]

1.3 TESTING AND METRICS

Various numbers of testing strategies have been devised to deliver the code to the customer. Metrics can be classified into process or code metrics. The number of products can be in any form; for example, it can be mobile application software or web development software codes. Here we, in this paper for testing any such program code, have used various simulations on the lines of code.

1.4 STATISTICAL ANALYTICS BASED MODULES

1.4.1 TEST CASE PRODUCTIVITY

The formulation formulated the test case writing productivity on the basis of which a conclusive remark can be as follows:

$$\text{Test case productivity} = \left[\frac{\text{Total raw test steps}}{\text{Efforts (hours)}} \right] \text{Step(s)/h} \quad (1.4)$$

For example, effort taken for writing 183 steps is around 8 h. Therefore, the test case productivity (TCP) is equals 22.8 when 183 steps divided by 8 h. The rounded value of test case productivity = 23 steps/h. The value of TCP can be compared with the preceding release(s) and successful closure can be drawn.

1.4.2 TEST EXECUTION SUMMARY

This metric or formulation classifies test cases their condition and motivation for a variety of test cases and gives the cavernous condition of release available when needed. It also classifies reasons if not able to test and a test case fails. Data can be collected for the number of test cases carried out with the status like:

- Pass/fail and reason for failure.
- Time crisis, postponed defect, arrangement issue, out of scope.
- Unable to test with reason (Fig. 1.1).

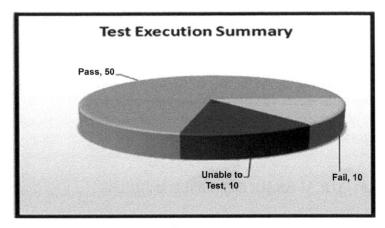

FIGURE 1.1   Test execution summary.[12]

1.4.3   DEFECT ACCEPTANCE (DA)

The formulation concludes the number of valid defects that testing team can identify during execution and is achieved after comparing to the previous release for getting better outputs.

$$\text{Defect acceptance} = \left[\frac{\text{Number of valid defects}}{\text{Total number of defects}} \times 100 \right] \% \qquad (1.5)$$

1.4.4   DEFECT REJECTION

The formulation determines the number of defects discarded during execution and offer percentage of unacceptable defect opened by testing team and can be controlled.

$$\text{Defect Rejection} = \left[\frac{\text{Number of Defect(s) Rejected}}{\text{Total number of defects}} \times 100 \right] \% \qquad (1.6)$$

1.4.5   BAD FIX DEFECT (B)

Whenever a defect is resolved and in return gives birth to new defect(s) is a bad fix defect. The formulation determines how effective defect resolution process and results percentage of bad defect to be resolved and forbidden.

$$\text{Bad fix defect} = \left[\frac{\text{Number of bad fix defect(s)}}{\text{Total number of valid defects}} \times 100 \right] \% \qquad (1.7)$$

1.4.6   TEST EXECUTION PRODUCTIVITY/ANOMALY ANALYSIS

The formulation provides the execution of test cases and their yield which can in advance be analyzed to get a conclusive result.

$$\text{Test execution productivity} = \left[\frac{\text{Total no. of TC executed (Te)}}{\text{Execution efforts (hours)}} \times 8 \right] \text{Execution(s)/Day} \qquad (1.8)$$

where Te is calculated as:

$$\text{Te} = \text{Base test case} + ((T(0.33) \times 0.33) + (T(0.66) \times 0.66) + (T(1) \times 1))$$

where
   Base test case = Number of TC executed atleast once.
   $T(1)$ = Number of TC retested with 71%–100% of total test case layers.
   $T(0.66)$ = Number of TC retested with 41%–70% of total test case layers.
   $T(0.33)$ = Number of TC retested with 1%–40% of total test case layers.

1.4.7   EFFORT VARIANCE

The formulation presents the variance in estimation of the effort (Fig. 1.2).

$$\text{Effort variance} = \left[\frac{\text{Actual effort} - \text{Estimated effort}}{\text{Estimated effort}} \times 100 \right] \% \qquad (1.9)$$

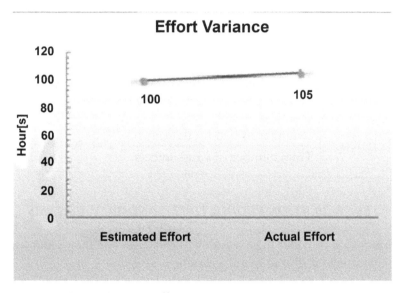

FIGURE 1.2 Effort variance trend.[14]

1.4.8 SCHEDULE VARIANCE

These formulations tell about any variation in the schedule that is estimated earlier and it will give output as the number of days (Fig. 1.3).

$$\text{Schedule variance} = \left[\frac{\text{Actual no. of days} - \text{Estimated no. of days}}{\text{Estimated no. of days}} \times 100 \right] \% \qquad (1.10)$$

1.4.9 SCOPE CHANGE

This formulation tells about the stability of the testing scope.

$$\text{Scope change} = \left[\frac{\text{Total scope} - \text{Previous scope}}{\text{Previous scope}} \times 100 \right] \% \qquad (1.11)$$

Here
Total scope = Previous scope + New scope if and only if scope > increases.
Total scope = Previous scope − New scope if and only if scope < decreases.

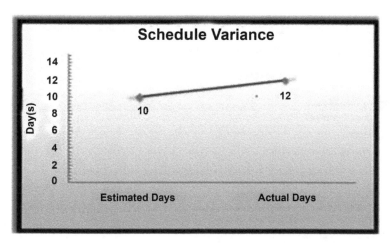

FIGURE 1.3   Schedule variance.[15]

1.5   PROPOSED APPROACH USING SIMULATION SCENARIO-1

In the scenario-I, the following figure shows the simulator implementation and lines of code written to test the accuracy of the traditional as well as Halstead approach (Figs. 1.4 and 1.5).

FIGURE 1.4   Screenshot of the implementation of simulator.[16]

Stats Object
(
distinct operators → 5
operators → 10
distinct unique operands variables constants → 3
number of operands variables constants → 10
number of struct used 0

number of classes → 0
number of constructors destructors → 0
lines of code → 21
comment lines → 5
friend functions → 0
virtual functions → 0
file pointers → 0
)

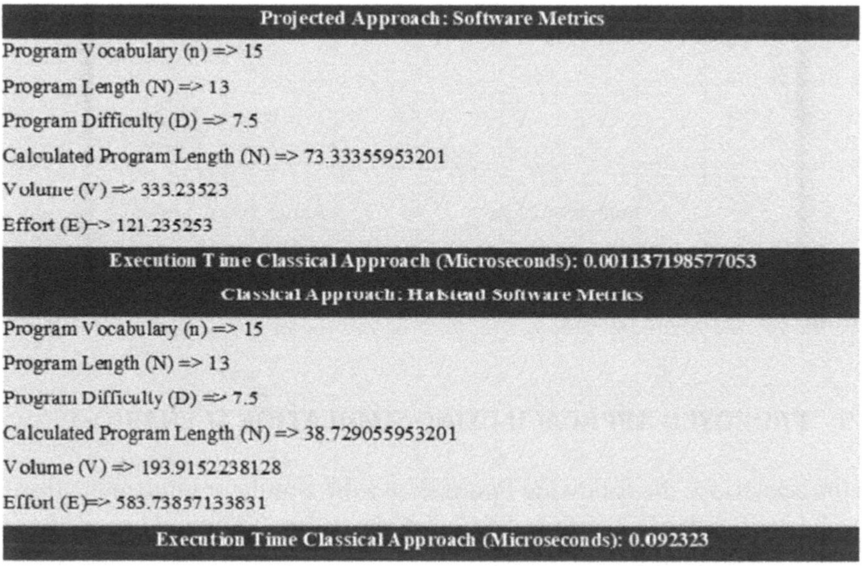

Projected Approach: Software Metrics
Program Vocabulary (n) => 15
Program Length (N) => 13
Program Difficulty (D) => 7.5
Calculated Program Length (N) => 73.33355953201
Volume (V) => 333.23523
Effort (E) => 121.235253
Execution Time Classical Approach (Microseconds): 0.001137198577053
Classical Approach: Halstead Software Metrics
Program Vocabulary (n) => 15
Program Length (N) => 13
Program Difficulty (D) => 7.5
Calculated Program Length (N) => 38.729055953201
Volume (V) => 193.9152238128
Effort (E) => 583.73857133831
Execution Time Classical Approach (Microseconds): 0.092323

FIGURE 1.5 Comparison of the classical and projected approach on accuracy in the same scenario.[17]

1.5.1 ADVANTAGES OF THE PROPOSED IMPLEMENTED WORK

As it is evident from the simulation results that the comparison shows it takes:

- Less execution time.
- Less complex.
- More reliable in terms of program length and efforts parameters (Fig. 1.6).

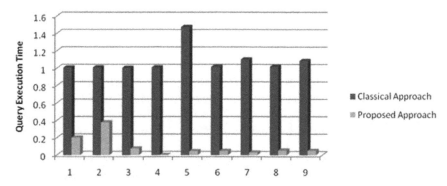

FIGURE 1.6   Implementation time analysis over classical and proposed approach.

1.6   IMPLEMENTATION/SIMULATION SCENARIO-2

In Table 1.1, it is evident that the second scenario is created with the number of lines is 10 LOC run or executed on live server, and accuracy is achieved 0.234234 s over 1.323412 s overproposed and classical approaches, respectively.

TABLE 1.1   Accuracy Over Classical and Proposed Approach.

Number of lines	Classical approach	Proposed approached
10	1.323412	0.234234

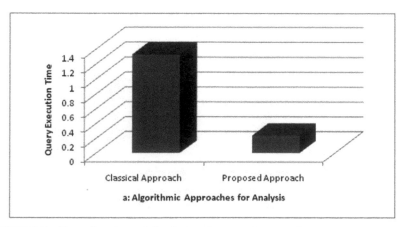

FIGURE 1.7   Execution time of classical and proposed approach.

Figure 1.7 also shows the status of the execution time when square bar charts between classical and proposed approach are analyzed.

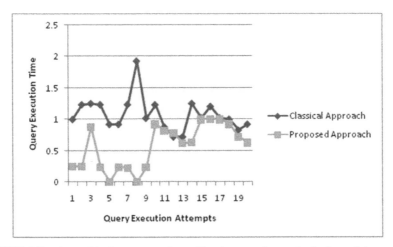

FIGURE 1.8   A graphical representation of implementation analysis time of the queries over classical versus projected.

From Figure 1.8, it is awfully apparent from the simulations that the projected algorithmic implementation is comparatively very lesser from simulations on the classical method in each execution stab when the graph is plotted between query execution time and query execution attempts.

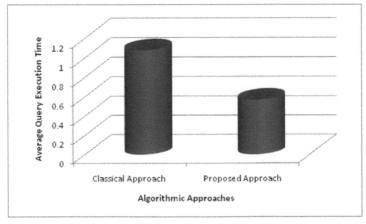

FIGURE 1.9   A bar-based comparison & execution average time analysis of queries from the classical and projected algorithmic implementation.

Figure 1.9 shows that the bar-based comparison chart is built and shows that projected algorithmic implementation in comparison to the classical approach is winning less time in execution. The situation taken is average of 20 lines of code that were executed on a live server and enticing results from the distributed databases.

1.7 CONCLUSION AND FUTURE SCOPE

In the work, the key focus is on the projected algorithmic implementation factor of presentation and is possible to top competence altitude. After we have analyzed the results fetched from the server, the proposed system is found better as compared to the classical approach. The fact has been established as the conclusion that the projected algorithmic implementation is to the point of top-level efficiency. There exist colossal free and open-source devices for execution testing and security review, it is encouraged to utilize numerous apparatuses for testing of the same application with the goal that the consistency of the application just as the testing device can be assessed. In the work, the key spotlight is on the projected algorithmic implementation as the boundary of execution and made it conceivable to the top proficiency level. When we have investigated the outcomes gotten from the worker, the proposed framework is discovered better when contrasted with the traditional methodology. The projected algorithmic implementation is to the point of high-level proficiency, and the old-style approach is exceptionally a long way presentation as reality is being built. The graphical portrayal indicating the presentation intensity of two methodologies: projected algorithmic implementation is on the higher piece of exhibition when contrasted with the old style. The metaheuristic-based usage can be played out that incorporates subterranean insect state enhancement, bumblebee calculation, mimicked toughening, and numerous assorted others. Such algorithmic methodology ought to give better outcomes when we move toward metaheuristics. This examination work for the most part talks about Halstead programming multifaceted nature measurements for explicit programming dialects.

KEYWORDS

- **software quality**
- **software code evaluation**
- **software quality assurance**
- **software risk mitigation**
- **software risk analysis**

REFERENCES

1. Dalla Palma, S.; Nucci, D. D.; Palomba, F.; Tamburri, D. A. Towards a Catalogue of Software Quality Metrics for Infrastructure Code. *J. Syst. Softw.* **2020,** 110726.

2. Yadav, S.; Kishan, B. Analysis and Assessment of Existing Software Quality Models to Predict the Reliability of Component-Based Software. *Int. J. Emerg. Trends Eng. Res.* **2020,** *8* (6).

3. Spinellis, D. *Code Quality: The Open Source Perspective*; Adobe Press, 2006.

4. McIntosh, S.; Kamei, Y.; Adams, B.; Hassan, A. E. An Empirical Study of the Impact of Modern Code Review Practices on Software Quality. *Empir. Softw. Eng.* **2016,** *21* (5), 2146–2189.

5. Karout, R.; Awasthi, A. Improving Software Quality Using Six Sigma DMAIC-Based Approach: A Case Study. *Busi. Process Manage. J.* **2017.**

6. Perera, P.; Silva, R.; Perera, I. Improve Software Quality through Practicing DevOps. In *2017 Seventeenth International Conference on Advances in ICT for Emerging Regions (ICTer)*; IEEE, Sept 2017; pp 1–6.

7. Rempel, P.; Mäder, P. Preventing Defects: The Impact of Requirements Traceability Completeness on Software Quality. *IEEE Trans. Softw. Eng.* **2016,** *43* (8), 777–797.

8. Behnamghader, P.; Alfayez, R.; Srisopha, K.; Boehm, B. Towards Better Understanding of Software Quality Evolution through Commit-Impact Analysis. In *2017 IEEE International Conference on Software Quality, Reliability and Security (QRS)*; IEEE, 2017; pp 251–262.

9. Siavvas, M.; G.; Chatzidimitriou, K. C.; Symeonidis, A. L. QATCH-An Adaptive Framework for Software Product Quality Assessment. *Expert Syst. App.* **2017,** *86,* 350–366.

10. Spadini, D.; Palomba, F.; Zaidman, A.; Bruntink, M.; Bacchelli, A. On the Relation of Test Smells to Software Code Quality. In *2018 IEEE International Conference on Software Maintenance and Evolution (ICSME)*; IEEE, 2018; pp 1–12.

11. Adewumi, A.; Misra, S.; Omoregbe, N.; Crawford, B.; Soto, R. A Systematic Literature Review of Open Source Software Quality Assessment Models. *Springer Plus* **2016,** *5* (1), 1–13.

12. Nistala, P.; Nori, K. V.; Reddy, R. Software Quality Models: A Systematic Mapping Study. In *2019 IEEE/ACM International Conference on Software and System Processes (ICSSP)*; IEEE, 2019; pp 125–134.

13. Carrozza, G.; Pietrantuono, R.; Russo, S. A Software Quality Framework for Large-Scale Mission-Critical Systems Engineering. *Info. Softw. Technol.* **2018,** *102,* 100–116.

14. Huang, J.; Keung, J. W.; Sarro, F.; Li, Y-F.; Yu, Y-T.; Chan, W. K.; Sun. H. Cross-Validation Based K Nearest Neighbor Imputation for Software Quality Datasets: An Empirical Study. *J. Syst. Softw.* **2017,** *132,* 226-252.

15. Foidl, H.; Felderer, M. Integrating Software Quality Models Into Risk-Based Testing. *Softw. Qual. J.* **2018,** *26* (2) (2018), 809–847.

16. García-Mireles, Gabriel Alberto, Mª Ángeles Moraga, Félix García, Coral Calero, and Mario Piattini. Interactions between Environmental Sustainability Goals and Software Product Quality: A Mapping Study. *Info. Softw. Technol.* **2018,** *95,* 108–129.

17. Russo, D.; Ciancarini, P.; Falasconi, T.; Tomasi, M. Software Quality Concerns in the Italian Bank Sector: The Emergence of a Meta-Quality Dimension. In *2017 IEEE/ ACM 39th International Conference on Software Engineering: Software Engineering in Practice Track (ICSE-SEIP)*; IEEE, 2017; pp 63–72.

18. Akbar, M. A.; Sang, J.; Khan, A. A.; Shafiq, M.; Hussain, S.; Hu, H.; Elahi, M.; Xiang, H. Improving the Quality of Software Development Process by Introducing a New Methodology–AZ-Model. *IEEE Access* **2017,** *6,* 4811–4823.

19. Mittal, M.; Sharma, R. K.; Singh, V. P. Validation of k-Means and Threshold Based Clustering Method. *Int. J. Adv. Technol.* **2014,** *5* (2), 153–160.

20. Mittal, M.; Balas, V. E.; Goyal, L. M.; Kumar, R., Eds. *Big Data Processing Using Spark in Cloud*; Springer, 2019.

21. Saxena, A.; Goyal, L. M.; Mittal, M. Comparative Analysis of Clustering Methods. *Int. J. Comput. App.* **2015,** *118* (21).

22. Aggarwal, A.; Aggarwal, A. Design of Software Process Improvement Model. *Int. J. Comput. App.* **2012,** 49 (13).

23. Aggarwal, A. An Effective Model for Software Risk Estimation, 2014.

24. Gakhar, S.; Agarwal, A. Team Based Refactoring of Object Oriented System, 2015.

25. Escalated Methods for Software Defects Audit in Repercussion and Effects Construe to Nature Inspired and Behavior Driven Mechanisms. *Regular Issue* Aug 30, **2019a,** *8* (6), 1779–1783. DOI:10.35940/ijeat.f8442.088619.

26. Usage Patterns and Implementation of Random Forest Methods for Software Risk and Bugs Predictions. *Special Issue* Aug 23, **2019b,** *8,* no. 9S, 927–932. DOI:10.35940/ ijitee.i1150.0789s19.

27. Aggarwal, A.; Dhindsa, K. S.; Suri, P. K. An Empirical Evaluation of Assorted Risk Management Models and Frameworks in Software Development. *Int. J. Appl. Evol. Comput. (IJAEC)* **2020a,** *11* (1), 52–62.

28. Aggarwal, A.; Dhindsa, K. S.; Suri, P. K. Design for Software Risk Management Using Soft Computing and Simulated Biological Approach. *Int. J. Sec. Priv. Pervasive Comput. (IJSPPC)* **2020b,** *12* (2), 44–54.

29. Aggarwal, A.; Dhindsa, K. S.; Suri, P. K. A Pragmatic Assessment of Approaches and Paradigms in Software Risk Management Frameworks. *Int. J. Nat. Comput. Res. (IJNCR)* **2020c,** *9* (1), 13–26.

30. Aggarwal, A.; Dhindsa, K. S.; Suri, P. K. Performance-Aware Approach for Software Risk Management Using Random Forest Algorithm. *Int. J. Softw. Innov. (IJSI)* **2021,** *9* (1), 12–19.

CHAPTER 2

Open Challenges in Software Measurements Using Machine Learning Techniques

SOMYA GOYAL*

*Department of Computer and Communication Engineering,
Manipal University Jaipur, Jaipur, India*

*Department of Computer Science and Engineering,
Guru Jambheshwar University of Science & Technology, Hisar, India*

E-mail: somyagoyal1988@gmail.com

ABSTRACT

Software measurement (SM) is a highly popular research field in software engineering. It revolves around either measuring the quality or measuring the quantity of a software and its attributes. The qualitative measurement of software has evolved as a research field software defect prediction (SDP). SDP highlights the modules that are prone to errors and may cause a failure of future software products. SDP allows focusing on the testing efforts on the predicted faulty modules and reduces the testing time and cost. Machine learning algorithms have found a fertile ground in the SDP field and yielded so many variant algorithms for effective SDP. Quantitative measurement has evolved as a software effort estimation problem. Since the past three decades, ML has been prominently deployed in SM. This work is dedicated to the identification of the open challenges

Computational Intelligence Applications for Software Engineering Problems. Parma Nand, PhD, Rakesh Nitin, PhD, Arun Prakash Agrawal, PhD & Vishal Jain, PhD (Eds.)
© 2023 Apple Academic Press, Inc. Co-published with CRC Press (Taylor & Francis)

in SMs using machine learning techniques and the remedies proposed in the literature. This chapter reviews the current state in SM using ML to discover the open challenges. Later, the challenges are elaborated and possible remedies from the literature are discussed. The detailed insight into each challenge is offered by giving suitable case studies from real-life experiments. With each case study, the proposed solution is also offered. Later the chapter is concluded with special remarks on future work.

2.1   INTRODUCTION

Software measurement (SM) is the most important activity during software development. It is an ongoing umbrella activity. It involves the measurement of various metrics at different marked milestones during the engineering of the software product. The measurement plays two major roles: (1) to gauge the quality of to be delivered product and (2) to assess the effort requirements for building the product in advance.[1] High quality is ensured with software defect predictors by highlighting those software modules that require most of the testing efforts. It allows zero defect in the testing on those fault-prone modules.[2–5] The effort requirement is estimated using some regressor on previous project data.[6–7] In both cases, machine learning is a panacea. The attributes are collected from the previously completed projects, next the ML-based predictor predicts if it is fault-prone or not for each module and then the ML-based regressor computes the cost-estimation for new projects.

Since the past three decades, machine learning has replaced almost all the statistical and traditional algorithmic models. Researchers have proposed a plethora of ML-based model for software defect prediction (SDP) and software effort estimation (SEE). The backbone of the ML-based model is training the predictor or regressor on previous data and using the trained model for upcoming projects. While working on SDP and SEE, researchers have reported many problems and attempted to resolve them in numerous ways. But some problems are still open as of either left unanswered or the proposed solutions are not accepted universally.

This chapter contributes to highlight those issues and their remedies from literature. This chapter is organized as follows. Section 2.2 introduces the major issues in SDP and SEE and discusses the proposed remedies from the literature. Section 2.3 concludes the chapter with special remarks on the future scope of the work.

2.2 ISSUES IN SOFTWARE MEASUREMENTS

This section covers open issues in the research field of SDP and SEE as reported by researchers in literature. The proposed solutions are also discussed with case studies in this section.

2.2.1 IDENTIFICATION OF SUBSTANTIAL ATTRIBUTES

Software quality prediction and SEE need holistic view of the project in terms of measured attributes (or independent features) that are guiding factors for the dependent entity (defect/effort). It is not the case where we consider one or two attributes, but it is the case of considering all essential and substantial attributes. The selected attribute-set must be most significant, relevant, and nonredundant.[9,10] The most important aspect is that the selected generalized attribute subset must be highly correlated to the target entity and least correlated with other independent attributes. Variety of metrics[11,12] is being used in literature like object-oriented, static code metric, size metrics, etc.

The identification of the most significant features from all the available features is highly crucial for accurate SDP[9,13] and SEE.[14–16] It is an important task under data preprocessing to handle the "curse of dimensionality." It involves removal of irrelevant and redundant features to improve the accuracy of SDP or SEE model. The most popular techniques are filter methods, embedded methods, and wrapper methods.[17] The researchers suggest a variety of feature selection techniques and search-based techniques are getting boom. Some short case studies are as follows.

Moosavi et al.[18] used function point metrics and selected attributes from ISBSG dataset for effort estimation of a new project. They deployed Satin Bower Bird algorithm and ANFIS to compute the MMRE and PRED for the model. They advocated the use of selected function point metrics and selected features for accurate estimation.

Nam et al.[19] deeply studied the feature selection and proposed approach for defect prediction using a variety of metrics. They used statistical procedures and evaluated the accuracy of model. They have shown that their approach is better than competitive approaches.

Hosseini et al.[20] deployed search-based feature selection. They used CK metrics and LOC metrics from PROMISE dataset. They proposed a novel genetic search algorithm to select significant features.

Some studies advocate hybrid methods and some others suggest individual filter methods, but the answers are neither generalized nor univocal solutions for feature selection. Hence, the identification of significant metrics and features for the training of prediction or estimation model is still an open problem.

2.2.2   HANDLING THE INCONSISTENCIES IN DATASET

Another major issue is to deal with the inconsistencies in the datasets. The foundation of SMs using machine learning is the availability of enough training dataset. In order to accurately predict the defective modules and estimate the accurate cost of project, the developer needs to train the ML prediction (or estimation) model with enough good quality cross-project dataset. The noise or skewness in the dataset can lead to improper classification results and inaccurate regression results. The accuracy of the predictor is directly proportional to the quality of dataset. The public datasets are most widely used in the research field. Such datasets suffer from poor quality issues like the presence of outliers or noisy information and class-imbalance nature of data or skewness of data. Such issues hinder the successful prediction of quality and effort requirements.

2.2.2.1   OUTLIERS PROBLEM IN SOFTWARE EFFORT ESTIMATION

The existence of outliers in the training dataset hinders the accurate performance of the machine learning model, increases the training time by misleading the training process, and results in less accurate results. Outliers are the exceptional cases in the dataset. These instances are far away from the rest of the project data. Machine learning algorithms get negatively affected by the existence of outliers due to their sensitivity toward the distribution of project data and attributes. Some common methods to detect the outliers are box-plot method, Z-score method, and interquartile range method.[21] Literature suggests two approaches to handle outliers in SEE measurement—(1) data level and (2) algorithmic level. At the data level, the outliners can be handled in numerous ways like these can be dropped, capped to some thresholds, replaced by some other value, or normalization can be applied. At the algorithmic level, ML algorithms and evaluation criteria are selected which are robust to outliers. Literature reminds a wide range of opinions for outliers handling. Two short case studies are discussed here:.

Minku et al.[22] advocate data level approach to handle the existence of outliers. They have used *k*-means-based filtering to detect and remove the outliers in SEE. They have showcased that their approach results better values for MAE and PRED.

Mustapha et al.[23] advocate algorithm based approach to improve the performance of ML model in existence of outliers. They deployed random forest and proved that ensembles are better to handle outliers than the normal regression-based models for effort estimation. They performed experiments on both the most popular dataset for SEE, namely ISBSG and COCOMO.

Many researchers have contributed multiple ways to handle outliers in dataset for SEE, but none of the approach is univocal. Some suggest filtering at data preprocessing is best while others suggest algorithm must be altered to deal with outliers. It is still an open issue in the field of SMs using machine learning techniques.

2.2.2.2 *CLASS-IMBALANCE PROBLEM OR SKEWED DATASET IN DEFECT PREDICTION*

Imbalanced distribution in dataset raises the issues in training the classifiers. It occurs when the number of samples of individual classes is unequal and varies to a large extent. It is amplified if the minority class is of importance. In SDP, where the minority class represents fault-prone modules that is of high importance, class imbalance majorly impacts the performance of predictor. Due to class imbalance predictor gets biased toward the majority class and results in false positives, which is alarming in SDP. Literature gives a variety of methods to cope up with class imbalance. One most common method is sampling which may be undersampling in the case of availability of large datasets and oversampling in the case of nonavailability of large dataset. Some researchers have advocated ensembles as being robust to class-imbalance nature of training dataset. Few contributed to the field of class imbalance handling in SDP with penalizing methods like cost-sensitive algorithms. Let us have a look into three case studies dealing with class imbalance issue in SMs field.

Feng et al.[24] favored the sampling of dataset to handle class imbalance issue. They suggested data level preprocessing is better than algorithmic level techniques. Specifically, they advocated oversampling rather than undersampling. A novel complexity based oversampling technique has been proposed by them and compared with four other competitive oversampling techniques over 23 datasets from PROMISE repository.

Goyal[25] advocated ensembles being robust to the class-imbalance issue. The researcher proposed stacked ensemble to quality prediction and compared it with the most renowned ML methods over five datasets from NASA repository.

Siers et al.[26] proposed cost-sensitive penalization to handle class-imbalance while measuring software quality. They proposed cost-sensitive forest algorithm and compared it with six state-of-the-art methods over six datasets from NASA repository.

From the literature, a range of methods has been contributed by the researchers in the field of SMs. Some advocate resampling, others suggest ensemble-based approaches, cost-sensitive methods, and choice of metrics[27] are also becoming fashionable. This issue has not been resolved in a generalized aspect. Hence it needs some univocal solution.

2.2.3 MISSING GENERAL FRAMEWORK

Another issue with this emerging research field of SMs using machine learning is that different contributors have deployed different models with variety of ML algorithms over a wide range of datasets. This field is 32 years old now with millions of research papers, journal articles, and books. But the paradox is that one approach is totally different from other in terms of the technique applied to the specific dataset. A glimpse is given in Table 2.1. Hence, it is a major problem that there is no standard framework to implement machine learning for SMs. Some case studies are discussed below.

Tong et al.[45] proposed SDAEsTSE framework for SDP. They used SDAE to extract features from the dataset and hence reduce the dimensions. TSE was used for addressing the class imbalance problem. Deep Network SDAE is optimized with fivefold cross-validation. TSE involves two stages: first is a training of bagging, RF, AdaBoost. Second is combining the three base ensembles using weighted average probability. The dataset used is NASA dataset for defect prediction. SDAEsTSE is deployed using MATLAB tool and compared with random forest, Bayesian network, neural network on the accuracy measures of F1-measure, AUC, and Mathew's. Further, the comparison is statistically tested using Wilcoxon-rank-sum test and Cliff's delta. The proposed model-SDAEs TSE outperformed all other techniques statistically.

TABLE 2.1   Glimpse of State-of-the-art ML Techniques and Dataset.

S.No.	Study	Contribution	Technique used	Dataset used
1.	[28]	Fault prediction	SOM	AR3, AR4, AR5
2.	[29]	Defect prediction	NN + BEE COLONY	NASA MDP,
3.	[30]	Effort estimation	Ensemble of adjustment variants analogy-based	Promise, Desharnais, Kermerer
4.	[31]	Fault prediction	ANFIS, ANN, SVM	PROMISE
5.	[9]	Fault prediction	Ensemble	NASA, PC1, PC4, MC1
6.	[32]	Software development efforts prediction	ANNs + PSO/PCA/GA	COCOMO, NASA, Albrecht
7.	[33]	Fault prediction	ANN, ANFIS	PROMISE
8.	[34]	Effort estimation	Ensembles	ISBSG
9.	[35]	Effort estimation	Neural networks (MLP, GRNN, RBFNN, CCNN)	ISBSG
10.	[36]	Effort estimation	Hybrid MLP=Dilation + erosion +Linear	Keremerer, Deshaarnais, Albrecht
11.	[37]	Effort estimation	Satin Bower Bird + ANFIS	Three Real Datasets
12.	[38]	Defect prediction	Multi objective cross version	PROMISE
13.	[39]	Enhancement effort	Support vector regression	ISBSG
14.	[40]	Effort estimation	Analogy + Differential evolution	Desharnais, NASA, Maxwell, Albrecht
15.	[41]	Defect prediction	Ensemble oversampling	Promise
16.	[42]	Defect predictor	Naïve Bayes + Info diffusion	10 OSS
17.	[43]	Defect predictor	Stacked ensembles	NASA MDP, PROMISE, AEEEM
18.	[44]	Effort estimation	Artificial neural network	Desharnais

Benala et al.[40] proposed a novel analogy-based technique and implemented it in this work using the differential algorithm features for effort estimation. To improve the accuracy of analogy-based estimation (ABE) techniques, differential evolution (DE) mutations are combined to propose a new technique DABE. It is compared with the state-of-the-art methods also. This work utilized five mutations of DE including DE/rand/1/bin, DE/best/1/bin, DE/rand/2/bin, DE/rand/2/best, DE/rand/1/bin, DE/rand-to-best/1/bin for feature weight optimization of similarity functions of ABE technique. Six datasets utilized are Desharnais, COCOMO, NASA 93, Maxwell, Albrecht, China. MRE, PRED(25), MMRE, EF, MAR, SA, cliff's Delta are the performance criteria. DABE is set-up and five mutations applied to get five variants of DABE as DABE-1,2,3,4,5. Using SA and Δ, the best is selected. That version is compared against Convolution-ABE, PSO-ABE, GA-ABE. It is found that DABE-3 is promising and outperforms other techniques statistically.

The calamity in this situation is that all proposed frameworks are good at SMs and also better than competitive methods; but totally different from each other. Hence, it is desirable to have a standard generic framework to measure software accurately.

2.2.4   NONAVAILABILITY OF STANDARDIZED EVALUATION PLATFORM

In the field of SMs using machine learning, researchers have contributed a wide range of ML models for SDP and SEE, and reported performances using the evaluation criteria of their individual choices. The selection of evaluation criteria is solely their individual choice. There is no standardization for the evaluation of model. It leads to inconsistency in the reported results. No two studies can be compared because they report the performance of their models in different aspects. It is clear from Table 2.2. It is evident that for SEE problems, MRE is commonly used and for SDP, AUC is most popular. But no standard set of evaluation criteria is available.

TABLE 2.2   Glimpse of the State-of-the-Art Evaluation Criteria.

Sr. No.	Study	Contribution	Evaluation criteria
1.	[28]	Fault prediction	Error Rate, FPR, FNR
2.	[29]	Defect prediction	FPR, FNR, AUC
3.	[30]	Effort estimation	AE, MRE, MMRE, PRED (l), MAE, SA
4.	[31]	Fault prediction	ROC, AUC
5.	[9]	Fault Prediction	AUC
6.	[32]	Software development efforts prediction	MMRE, PRED(l)
7.	[33]	Fault prediction	ROC, AUC
8.	[34]	Effort estimation	MAE
9.	[35]	Effort estimation	MAR
10.	[36]	Effort estimation	MMRE, PRED(l)
11.	[37]	Effort estimation	MMRE, PRED(l)
12.	[38]	Defect prediction	Mcost, Recall
13.	[39]	Enhancement effort	MAR, MdAR
14.	[40]	Effort estimation	AE, MRE, MMRE, PRED(l), EF, SA
15.	[41]	Defect prediction	AUC
16.	[42]	Defect predictor	Recall, Precision, F-measure
17.	[43]	Defect predictor	Recall, Precision, F-measure
18.	[44]	Effort estimation	MRE, MMRE, MdMRE

2.2.5   EXPENSES OF SOFTWARE MEASUREMENTS

The expense of carrying out SMs using machine learning has totally been ignored by most of the researchers. The researchers have contributed their work in the field of SDP and SEE for measuring the quality and quantity of software. The irony is that the work is highly useful in the software engineering domain but lacks the exploration of expense of making such predictions. The risk associated with misclassification and inaccurate estimation is very expensive. This aspect has been neglected by researchers in majority. In SDP, a wrong classification can result in failure of the software. If a faulty module is categorized as a nonfaulty, then false-negative responses can damage the entire software. On the other hand, if a clean module is classified as a buggy module, then this false positive case causes additional testing cost. Similarly, in SEE, if overestimation is done, then the resources will be wasted. In other situation, if underestimated then the project will suffer with lack of resources. In all cases, the expense of measuring software using machine learning is an important aspect needed to consider while proposing any work in this direction. Till date, it is not explored in full-fledge manner.

The above-mentioned issues are open to be explored and find solutions. In the next section, a few issues are being handled by the author in a novel approach.

2.3 CONCLUSION AND FUTURE SCOPE

SM is an essential activity during software development. It enables the development team to assess the resource requirement in advance, to focus the testing efforts on the faulty modules, and to ensure the timely delivery of the high-quality product. This chapter brought into the notice some aspects that are either unexplored or partially explored. These issues are needed to be explored to find some standardized generic solution.

This chapter provides a guiding torch for the upcoming researchers or contributors to assess the flaws in SM. It presents the prevailing flaws in an aggregated fashion. In future, this work can be extended to solve the other open issues in SM using machine learning.

KEYWORDS

- **software measurement**
- **software defect prediction**
- **software effort estimation**
- **machine learning**

REFERENCES

1. Goyal, S.; Bhatia, P. K. Empirical Software Measurements with Machine Learning. In *Computational Intelligence Techniques and Their Applications to Software*

Engineering Problems; Bansal, A., Jain, A., Jain, S., Jain, V., Choudhary, A., Eds.; Boca Raton: CRC Press, 2021; pp 49–64. https://doi.org/10.1201/9781003079996

2. Özakıncı, R.; Tarhan, A. Early Software Defect Prediction: A Systematic Map and Review. *J. Syst. Softw.* **2018,** *144,* 216–239.

3. Malhotra, R. A Systematic Review of Machine Learning Techniques for Software Fault Prediction. *Appl. Soft Comput.* **2015,** *27,* 504–518.

4. Goyal, S.; Bhatia, P. K. Software Quality Prediction Using Machine Learning Techniques. In *Innovations in Computational Intelligence and Computer Vision. Advances in Intelligent Systems and Computing*; Sharma, M. K., Dhaka, V. S., Perumal, T., Dey, N., Tavares, J. M. R. S., Eds., Vol. 1189; Springer: Singapore, 2021; pp 551–560. https://doi.org/10.1007/978-981-15-6067-5_62

5. Goyal, S.; Bhatia, P. K. Comparison of Machine Learning Techniques for Software Quality Prediction. *Int. J. Knowledge Syst. Sci. (IJKSS)* 2020, *11* (2), 21–40. DOI: 10.4018/IJKSS.2020040102

6. Kitchenham, B.; Brereton, O. P.; Budgen, D.; Turner, M.; Bailey, J.; Linkman, S. Systematic Literature Reviews in Software Engineering–a Systematic Literature Review. *Info. Softw. Technol.* **2009,** *51* (1), 7–15.

7. Wen, J.; Li, S.; Lin, Z.; Hu, Y.; Huang, C. Systematic Literature Review of Machine Learning Based Software Development Effort Estimation Models. *Info. Softw. Technol.* **2012,** *54* (1), 41–59.

8. Goyal, S.; Parashar, A. Machine Learning Application to Improve COCOMO Model using Neural Networks, *International Journal of Information Technology and Computer Science (IJITCS-2018)* Mar **2018,** *10* (3), 35–51. DOI: 10.5815/ijitcs.2018.03.05

9. Laradji, I. H., Alshayeb, M.; Ghouti, L. Software Defect Prediction Using Ensemble Learning on Selected Features. *Info. Softw. Technol.* **2015,** *58,* 388–402.

10. Oliveira, A. L.; Braga, P. L.; Lima, R. M.; Cornélio, M. L. GA-Based Method for Feature Selection and Parameters Optimization for Machine Learning Regression Applied to Software Effort Estimation. *Info. Softw. Technol.* **2010,** *52* (11), 1155–1166.

11. Arar, Ö.F.; Ayan, K. A Feature Dependent Naive Bayes Approach and Its Application to the Software Defect Prediction Problem. *Appl. Soft Comput* **2017,** *59,* 197–209.

12. Kumar, P. S.; Behera, H. S.; Kumari, A.; Nayak, J.; Naik, B. Advancement from Neural Networks to Deep Learning in Software Effort Estimation: Perspective of Two Decades. *Comput. Sci. Rev.* **2020,** *38,* 100288.

13. Catal, C.; Diri, B. Investigating the Effect of Dataset Size, Metrics Sets, and Feature Selection Techniques on Software Fault Prediction Problem. *Info. Sci.* **2009,** *179* (8), 1040–1058.

14. Sehra, S. K.; Brar, Y. S.; Kaur, N.; Sehra, S. S. Research Patterns and Trends in Software Effort Estimation. *Info. Softw. Technol.* **2017,** *91,* 1–21.

15. Goyal, S.; Bhatia, P. K. Feature Selection Technique for Effective Software Effort Estimation Using Multi-Layer Perceptrons. In *Proceedings of ICETIT 2019. Lecture Notes in Electrical Engineering*, Vol 605; Springer: Cham, 2020; pp 183–194. https://doi.org/10.1007/978-3-030-30577-2_15

16. Goyal, S.; Bhatia, P. K. GA Based Dimensionality Reduction for Effective Software Effort Estimation Using ANN. *Adv. App. Math. Sci.* June **2019,** *18* (8), 637–649.

17. John, G. H.; Kohavi, R.; Pfleger, K. Irrelevant Features and the Subset Selection Problem. In *Machine Learning Proceedings*; Elsevier, 1994; pp 121–129.

18. Moosavi, S. H. S.; Bardsiri, V. K. Satin Bowerbird Optimizer: A New Optimization Algorithm to Optimize ANFIS for Software Development Effort Estimation. *Eng. App. Artif. Intell.* **2017**, *60*, 1–15.

19. Nam, J.; Fu, W.; Kim, S.; Menzies, T.; Tan, L. Heterogeneous Defect Prediction. *IEEE Trans. Softw. Eng.* **2017**, *44* (9), 874–896.

20. Hosseini, S.; Turhan, B.; Mäntylä, M. A Benchmark Study on the Effectiveness of Search-Based Data Selection and Feature Selection for Cross Project Defect Prediction. *Info. Softw. Technol.* **2018**, *95*,.296–312.

21. Zhu, J.; Ge, Z.; Song, Z.; Gao, F. Review and Big Data Perspectives on Robust Data Mining Approaches for Industrial Process Modeling with Outliers and Missing Data. *Annu. Rev. Control* **2018**, *46*, 107–133.

22. Minku, L.L.; Yao, X. Ensembles and Locality: Insight on Improving Software Effort Estimation. *Info. Softw. Technol.* **2013**, *55* (8), 1512–1528.

23. Mustapha, H.; Abdelwahed, N. Investigating the Use of Random Forest in Software Effort Estimation. *Procedia Comput. Sci.* **2019**, *148*, 343–352.

24. Feng, S.; Keung, J.; Yu, X.; Xiao, Y.; Bennin, K.E.; Kabir, M.A.; Zhang, M. COSTE: Complexity-based OverSampling TEchnique to Alleviate the Class Imbalance Problem in Software Defect Prediction. *Info. Softw. Technol.* **2021**, *129*, 106432.

25. Goyal, S. Heterogeneous Stacked Ensemble Classifier for Software Defect Prediction. In *2020 Sixth International Conference on Parallel, Distributed and Grid Computing (PDGC)*; Waknaghat: Solan, India, 2020; pp 126–130. DOI: 10.1109/PDGC50313.2020.9315754.

26. Siers, M. J.; Islam, M. Z. Software Defect Prediction Using a Cost Sensitive Decision Forest and Voting, and a Potential Solution to the Class Imbalance Problem. *Info. Syst.* **2015**, *51*, 62–71.

27. Öztürk, M. M. Which Type of Metrics Are Useful to Deal with Class Imbalance in Software Defect Prediction? *Info. Softw. Technol.* **2017**, *92*, 17–29.

28. Abaei, G.; Selamat, A.; Fujita, H. An Empirical Study Based on Semi-Supervised Hybrid Self-Organizing Map for Software Fault Prediction. *Knowl. Based Syst.* **2015**, *74*, 28–39.

29. Arar, Ö.F.; Ayan, K. Software Defect Prediction Using Cost-Sensitive Neural Network. *Appl. Soft Comput. J.* **2015**, *33*, 263–277.

30. Azzeh, M.; Nassif, A. B.; Minku, L. L. An Empirical Evaluation of Ensemble Adjustment Methods for Analogy-Based Effort Estimation. *J. Syst. Softw.* **2015**, *103*, 36–52.

31. Erturk, E.; Sezer, E.A. A Comparison of Some Soft Computing Methods for Software Fault Prediction. *Expert Syst. Appl.* **2015**, *42*, 1872–1879.

32. Bisi, M.; Goyal, N. K. Software Development Efforts Prediction Using Artificial Neural Network. *IET Softw.* **2016**, 10 (3), 63–71.

33. Erturk, E.; Sezer, E. A. Iterative Software Fault Prediction with a Hybrid Approach. *Appl. Soft Comput.* **2016**, *49*, 1020–1033.

34. Idri, A.; Hosni, M.; Abran, A. Improved Estimation of Software Development Effort Using Classical and Fuzzy Analogy Ensembles. *Appl. Soft Comput.* **2016**, *49*, 990–1019.

35. Nassif, A. B.; Azzeh, M.; Capretz, L. F.; Ho, D. Neural Network Models for Software Development Effort Estimation: A Comparative Study. *Neural Comput. Appl.* Nov **2016,** 27 (8), 2369–2381.
36. Araujo, R. D. A.; Oliveira, A. L. I.; Meira, S. A Class of Hybrid Multilayer Perceptrons for Software Development Effort Estimation Problems. *J. Expert Syst. App.* Aug **2017.**
37. Moosavi, S. H. S.; Bardsiri, V. K. Satin Bowerbird Optimizer: A New Optimization Algorithm to Optimize ANFIS for Software Development Effort Estimation. *J. Eng. App. Artif. Intell.* Jan **2017,** *60,* 1-15.
38. Murillo-Morera, J.; Quesada-López, C.; Castro-Herrera, C.; Jenkins, M. A Genetic Algorithm Based Framework for Software Effort Prediction. *J. Softw. Eng. Res. Dev.* Dec **2017,** *5* (1), 4.
39. García-Floriano, A.; López-Martín, C.; Yáñez-Márquez, C.; Abran, A. Support Vector Regression for Predicting Software Enhancement Effort. *Info. Softw. Technol.* May **2018,** *97,* 99–109.
40. Benala, T. R.; Mall, R. DABE: Differential Evolution in Analogy-Based Software Development Effort Estimation. *Swarm Evol. Comput.* Feb **2018,** *38,* 158–172.
41. Huda, S.; Liu, K.; Abdelrazek, M.; Ibrahim, A.; Alyahya, S.; Al-Dossari, H.; Ahmad, S. An Ensemble Oversampling Model for Class Imbalance Problem in Software Defect Prediction. *IEEE Access* **2018,** *6,* 24184–24195. DOI: 10.1109/access.2018.2817572
42. Wu, Y.; Huang, S.; Ji, H.; Zheng, C.; Bai, C. A novel Bayes Defect Predictor Based on Information Diffusion Function. *Knowl. Based Syst.* **2018,** *144,* 1–8. DOI: 10.1016/j. knosys.2017.12.015
43. Li, Z.; Jing, X-Y.; Zhu, X.; Zhang, H.; Xu, B.; Ying, S. Heterogeneous Defect Prediction with Two-Stage Ensemble Learning. *Autom. Softw. Eng.* **2019,** *26,* 599. DOI: 10.1007/s10515-019-00259-1
44. Goyal, S.; Bhatia, P. K. A Non-Linear Technique for Effective Software Effort Estimation Using Multi-Layer Perceptrons, In 2019 International Conference on Machine Learning, Big Data, Cloud and Parallel Computing (COMITCon); Faridabad, India, 2019; pp 1–4. DOI: 10.1109/COMITCon.2019.8862256
45. Tong, H.; Liu, B.; Wang, S. Software Defect Prediction Using Stacked Denoising Autoencoders and Two-Stage Ensemble Learning. *Info. Softw. Technol.* Apr. **2018,** *96,* 94–111.

CHAPTER 3

Empirical Software Engineering and Its Challenges

SUJIT KUMAR[1*], SPANDANA GOWDA[1], and VIKRAMADITYA DAVE[2]

1Department of Electrical and Electronics Engineering, Jain (Deemed-to-be University), Bangalore 560041, Karnataka, India

2Department of Electrical Engineering, College of Technology and Engineering, Udaipur 313001, Rajasthan, India

*Corresponding author. E-mail: sujitvj.kumar@gmail.com

ABSTRACT

In the current years, "artificial intelligence (AI)" systems have shown encouraging results. Numerous of these milestones have remained accomplished in academic settings or by significant expertise firms through extremely trained study assemblies and specialized substructure funding. The development of superior yield set systems with AI mechanisms has proved stimulating aimed at corporations deprived of significant exploration collections or progressive substructure. There is a substantial shortage of healthy operative resources and the most exemplary practice for designing AI schemes. It aims to classify the critical tasks by smearing an explanatory study method closely associated with corporations of variable sizes and types. Related to additional parts like "software engineering (SE)" or "database technologies," it is flawless that AI remains motionless relatively juvenile and that extra effort is required to promote high-quality

Computational Intelligence Applications for Software Engineering Problems. Parma Nand, PhD, Rakesh Nitin, PhD, Arun Prakash Agrawal, PhD & Vishal Jain, PhD (Eds.)

systems. Problems described in this chapter can be used to direct upcoming studies in SE and AI societies. Together, we will encourage an incredible amount of corporations to start captivating the benefit of the high latent of AI tools.

3.1   INTRODUCTION

Presently, artificial intelligence (AI) is well known for its achievement in territories like "computer apparition" activities (e.g., target detection[1] and picture production[2]) utilizing neural convolution networks, regular linguistic comprehension employing "recurrent neural networks (RNN),"[3] and system tactical reasoning by "deep reinforcement learning (DRL)".[4] A significant transformation from the traditional approaches is that AI eventually acquires how to embody data through multiple abstraction levels.[5,6] Owing to the enormous amount of work that needs to be performed to create a text-based representation manually, this method can now be automated. The usage of an automated and data-driven system to recreate the functionality of the model, reducing the need for manual feature engineering.[7,8]

AI is made and constructed whereas software engineering (SE) is described and engineered into some of the most complicated and demanding systems engineers have produced. Tech engineers also do not earn all the respect they merit because their contributions are often viewed as behind-the-scenes, secret, or obscured behind code. Extra inspire with the "Eiffel tower," a masterpiece in French engineering reputedly possessing the sum of 2.5 million rivets. Tourists are awe-struck by the impressive technical prowess used to build The Eiffel Tower. In comparison, the technical challenges of application computing do not lead the public to consider hardware's viewpoint. The Eiffel Tower does not break down when one of the 2.5 million rivets split, but the entire tower itself is impacted.

The iPhone is quite high tech in that it incorporates an immense number of engineering code and far more code. Still, relative to an essential mobile phone, these variations are practically marginal, and the probability of all the code working at once is incredibly slim. As the space of inputs in an app on the phone (the phone's hardware) is likely to be greater than all the atoms in the measurable universe, but all but a single input may turn out to be inaccurate or null, it is difficult to even the tiniest app on the phone from its initial boot up from an app crash.[7]

In the face of the challenging task of planning, constructing, and evaluating engineering solutions for sizes, SE luckily takes crucial benefit that most engineers cannot possess; the SEs individual material (the software) may be used to outbreak the obstacles raised through the development of schemes in hardware. AI procedures are particularly well-tailored to this kind of complex programming work since they are programmed to overcome all the challenges that every employee would encounter when doing the same career-creating complicated software. What software developer (SD) does not like to assist intelligent software resources in his work?

This transparency and shared admiration between the technical fields of AI and SE has driven SDs to "innovate" within some of the approaches and systems that have arisen from the computational class of AI. It sees a wide variety of applications in SE. The most prominent of these are the ones that are so much narrower in reach and effect[8], and are

unique to the subject such as "Info Recovery (IR), Normal Linguistic Processing (NLP), and Machine Learning (ML)." It has been said that search-based information engineering is analogous to the world of analytical science and engineering. The first section talks about blurry and probabilistic strategies for thinking in the face of ambiguity. 3) Grouping, Learning, and Predictive Studies. As in every scientific sector, technological or otherwise, the field of AI is not merely an object waiting for the day of discovery. There have been many recent breakthroughs in AI, in one of which incredible goals that historically were not possible to tackle have been resolved. The prevailing effort has previously shown that there is substantial scope for SD to profit from AI methods. This chapter would include a short overview for emerging growth, outlining overall patterns, overlapping titles, access issues, and problems that have been studied so far.

Dissecting the world of AI from its subdiscipline into "softer" subdomains would maybe decrease the overall degree of analysis. However, as you will see, there is significant overlap between the science behind, or the e-phrase, of teaching and the science behind AI techniques, and thus this will be an error, although an enticing fault for persons whose certified life is expended through learning.

This chapter would quickly analyze three of the more common fields that AI dealt with SE. These subjects appear to concentrate on various areas, such as human–computer interaction. In conclusion, the chapter addresses the five problems that must be addressed for the creation of AI for SE.

3.2 AI USEFULNESS IN SE

The fields where AI strategies devise shown to be effective in SE study and preparation are described as "Probabilistic SE," "Cataloging, Knowledge and Estimating SE," and "Exploration Oriented SE." In "Fuzzy and probabilistic function," the goal is to extend to SE, AI techniques built to cope with real-world glitches that are, through its design, "fuzzy and probabilistic".[8] Presently logical match here, since more and more SE has to care for fuzzy, ill-defined, noisy, and imperfect knowledge, as its implementations move deeper into our messy, ambiguous, and imprecise lives. This is true not only of the software schemes we create but also of the mechanisms by which they are designed, all of which are focused on approximations.

A unique illustration of a probabilistic AI methodology proven to widely relevant in Information Manufacturing was the usage of "Bayesian probabilistic" reasoning for modeling software dependability,[5] one of the initial[6] instances of the acceptance of what could be considered "AI for SE," perhaps with hindsight. Another indication of the need for probabilistic thinking is the study of users, which, owing to the stochastic existence of human actions, necessarily involves an aspect of probability.[7] In work on classification, learning, and estimation, there was a great deal of attention in modeling and forecasting software charges as a fragment of scheme preparation. Like, a broad range of standard ML methods, like ANN, cased-based cognitive and rule-based inference, utilized for software scheme forecast,[8,9] ontology knowledge,[10] and imperfection calculation.[11]

A summary of ML approaches for information manufacturing can be contained in Menzies' work.[12] In the "Search-Based SE (SBSE)" position, the aim is to reformulate SE problems as optimization issues that can then be solved via computational exploration.[13,14] This has proven to be a generally available and effective method, with specifications and design presentations[15,16] for preservation and challenging.[17–19] Computational quest has been utilized for all technical backgrounds, not unbiased SE. The computer-generated existence of the program brands the perfect manufacturing content for computational exploration.[20] A current class offers a reference to SBSE.[21]

ML, particularly deep learning (DL), differs in part from conventional SE, whose activity is highly dependent on data from outside the world. Definitely, due to these conditions that ML is becoming practical. The main distinction between ML and non-ML schemes is that the data partially substitute the code in the ML framework. The knowledge procedure is used to inevitably find similarities in the statistics in its place of script firm implied laws. This means that the data can be checked as methodically as the encryption, then still a shortage of common experience about how to do so.

A considerable number of investigations were performed on the testing of software[11] and the testing of ML results. The connection between SE and ML has not been consequently deliberated.[12] It is not just the consistency of the model that needs to be checked but also the production-ready ML system.[13]

As DL is paired with the current development in big data, prospects and transformational possibilities arise for segments like banking, healthcare, industrial, and education.[14–18] Acting for "big data" introduces further technology service considerations. In its place just packing altogether statistics into reminiscence on a solitary computer, specific substructure could be required, such as "Apache Spark",[9] Apache Flank,[5] or "Google Dataflow".[11] Big data is also insecurely characterized by the "three Vs.: Volume, Variety, and Speed".[12] It is not just the amount of data that will involve large-scale data strategies, moreover the number of various data forms and formats. This is another power of DL that utilizes symbol knowledge to integrate multiple modes like pictures, acoustic, manuscript, and smooth statistics.[3,4] Though, this often raises difficulties about how to combine and convert various statistics bases.[5] Assumed the vast amount of multiple arrangements for pictures, audiovisual, manuscript, and tabular data, developing a data integration pipeline can be difficult. Besides, speed specifications (i.e., running period or in actual period minor dormancy requirements) can often include massive data strategies like flowing dispensation. This brings a range of complexities in developing an extensive statistics DL framework. It is not just the abstraction, transformation, and filling statistics process but also the use of modern dispersed exercise algorithms.

Applicable obligation, a symbol presented by "Ward Cunningham," may be used as an explanation for extended costs accrued by running rapidly to SE. It was claimed that ML networks take superior potential to accumulate technical debt.[8] ML programs encounter equal cipher obligations and the reliance on shifts in the exterior situation. Data dependences have been shown to create deficits comparable to encrypt requirements. There are currently few tools available for assessing and examining statistics dependences, remarkably static code analysis utilizing compilers and building systems.

An instance of a facts dependence instrument is defined by Poulding and Clark[9] where statistics foundations and topographies can be glossed. DL often enables specific models to be compiled from a collection of submodels and theoretically re-use pretrained strictures through so-named shift knowledge techniques. This introduces extra data addictions and exterior replicas that can be skilled independently and can also adjust conformation over the period. Exterior replicas incorporate additional dependence obligation and ownership costs for ML structures. Addiction commitment is described as significant sponsors to SE practical liability.[10]

3.3 AI ASSOCIATION METHODS WITH SE

It shows that all of the diverse habits in which AI methods utilized within SE yield precisely the same effects. Take examples such as those in probabilistic decision-making. The lines between these various processes are fuzzy, if not random, much like someone can effortlessly contemplate an estimation method as nonentity further possibility created by a rational. Think about a Bayesian model as a learner (a classifier) who also has to refine its likelihood predictions depending on its experiences.

Altogether the aspects where AI was applied to SE can be viewed as a way toward improvement whichever the manufacturing procedure or its goods. By itself, these are of Examine-Oriented SE. Furthermore, as we hear about our dilemma as one of finding the optimal solution in any given situation, or as one of seeking the most optimal solution with a given amount of cost over several conditions, we are still looking for the optimal path to a particular solution or a particular level of cost-benefit. Optimization activities typically take objective functions and conic constraints, which can also be formulated as observable targets and regulations that often exist in broad spaces that make ready for "computational search."

A similar interaction was found between ML methods toward computing and ML methods toward SE. ML is the learning of the use of algorithms to enhance the results of calculations. So as to recover, it is required a path to calculate our development and then can use it to improve our method conferring to that. Providentially, in SE scenarios, it may normally perform a significant range of applicant capacities to assess the efficacy of a proposed technological solution.

Similar to the work on ML and system boundary enforcement, there are several overlaps in work on genetic programming and researching the results of SBSE. In the past, one of the most commonly used computational search methods is "genetic programming",[10] employing thrilling current breakthroughs in automated "bug" adhesive,[7,9] it would be greatly beneficial because of porting among platforms, idioms, and software design standards[9] and sharing valuable and nonuseful possessions.[10]

And also, "genetic programming" can be spoken as well technique for studying patterns of SE. It can be used as a mirror from which it also appears as an ML method and an optimization strategy. SE essentially refers to finding out that there are incredibly close relationships between algorithms and computers and that there are also lessons to be gained from the way these relationships work. Similarly, SBSE is thinking about how to find select valuable insights in order to better than SE.

The demarcation of the issue of the way we interpret a given the word should not be talked over with such strength and complexity that it becomes the creation of an interminable and introspective direction of the built word. As compared to arbitrarily making neologisms with no tie to what came before, prior nomenclatures should be used as a basis for beginning new ones. For example, before any predictive algorithm can be fully executed, it must be configured using particular strategies.

The initial phase in the positive implementation of several AI methods to several SE problem fields is to catch an appropriate SE construction such that the AI technique becomes relevant. If the preparation is proficient, it usually unlocks a technical space of chance over many AI methods that will commercially move that repetitively proved in the exertion mentioned above.[13–17]

3.4   DESIGNATED TASKS

This segment list of briefly defined tasks at the node between AI and SE has been provided. It has been mapped into two classes: expansion and manufacturing tasks.

3.4.1   EXPANSION CHALLENGES

There are significant changes between the development of AI systems and conventional "SE systems." A big adjustment is that the data is used to program the machine as an alternative to manually writing the software code.[9] Planning for this project is challenging to do because of the complexity of data.

3.4.1.1   THOROUGH APPROACHES

Current SE architecture methods primarily rely on (1) the documentation, by way of test data, of a device under test; (2) the prediction of the quality of a tested system; or (3) the fitting of a specific equation to a tested system, without regard to the properties of the tested system itself. There is space to shift the chain of abstraction from one problem scenario into entire groups of problems, allowing it to establish methods for seeking a solution rather than a real solution.

Some initial work has already been done to search for resultant prospect deliveries for arithmetical challenges[9] and pathways in fertilization approaches for typical checks.[10] Work has also been carried out to search for program transformation tactics.[11,12]

Although that is real, this thesis remains based on terms. As far as looking for answers, whether it is by people or through teams, it needs to be watched if it will better wander from the quest for explanation occurrences to the hunt for strategy. In this study, we would be using the normal relations between brain imaging and ML, as the quest for methods can initially be conceived for learning method against preparation range. GP is renowned for its capacity to solve problems on which the result relies on a high number of variables. Instead of hoping to discover a fixed test input, why not gain insight into the relationship by utilizing a genetic-programming-type methodology to catch the following assessment contribution by evaluating the outcomes so far.[13]

As compared to a straightforward collection of specifications for the next version of the program, we should think of it as a significant measure toward the original strategic release preparation goals. The researcher must decide the release phase, the timing of the release, and identify the optimum release process based on the release history details.

3.4.1.2   EXPLOITING MULTICORE COMPUTING

Regarding AI, some view them as being quite computationally intensive and therefore not a strong candidate for solving the good kinds of large-scale problems confronting machine engineers. Some AI methods that might try to introduce to information manufacturing issues, like "evolutionary algorithms," "embarrassingly parallel" in their functionality, automatically pass into subcomputations that can be done in similar by one or more different activities. Parallelization option has been used in software remodeling,[14,15] idea location,[16] and reversion challenging.[17] While this exertion is hopeful, further effort is mandatory to completely leverage an immense possibility of the quickly rising amount of mainframes accessible.

Major critical tasks of multicore computing relics seek methods to convert current software design patterns hooked on logically parallel forms.[18] Without this, multicore execution will reduce performance because every essential is usually clocked at a low amount compared to a comparable single-core scheme.[19]

Most of the algorithms for SBSE in this chapter have a relationship to other algorithms that were developed before this novel. We have been able to establish that a cheap "General-Purpose-Graphics Processing Unit (GPGPU)" is capable of speeding up computations by as much as 25 times over a single count.[17] They also note that the scale-up in reaction to care is too high in the case of significant regression testing issues. The growing number of processor types allows for an exciting prospect of scalability, but we should be mindful of the possibility that our SE challenges may be scaling at the same pace as hardware.

3.4.1.3   ADVISING SOFTWARE DEVELOPERS

AI programs are structured to help correct software technical issues. Still, they are often used to provide vision to the areas of the software where the mistake is understood to be corrected and cannot be fixed due to the error being common knowledge.[20–25] Several facets of E-cigarette Technology continue to be checked and learned. For example, while much has been accomplished to define realistic specifications, project schedules, prototypes, process controls, etc., there is still a great deal of study that lets one understand the essence of the above problems.[26–28] For example, SBSE was utilized to expose tradeoffs between stakeholders (e.g., the specifications and the execution of the requirements) and between requirements and the implementation[29] and to analyze the aesthetics of applications[30] rigorously.

Of necessity, we know that work has already been performed on identifying the risks inherent in misestimating project specifications and finish times[31–33]—both of which, at the end of the day, are attempts at predictors of quality. The trend concerning these methods goes beyond the quest for a remedy for high-quality ventures; though they all rely on enhancing project forecasts by having some input.[34–36]

There are several fun and fascinating ways AI and ML methods can be used to obtain knowledge. Many scientific findings suggest neurocognitive problems of hooking people up to computers. Putting effort into finding a response to a problem is a lot more effort than developing a solution to a question.[37] This empirically driven study examines the effect of AI approaches on providing results rather than constructing and optimizing current proven best solutions that are inherently more challenging. The referees of such papers ought to do their bit by understating the difficulty of the task open. However, recent research from 2016 indicates AI is now at a stage where it outperforms most humans in the concentration of such functions, such as writing software programs.[38]

3.4.1.4   ACCUMULATION OF SMART OPTIMIZATION IN ARRANGED SOFTWARE

Much of the work on advancing AI for design has been performed in an alien AI off to enhance both the software method (like software development, design, and challenging) or the software itself (mechanically repairing, upgrading, and importing).[38–41] In mind this comes, if we can optimize the type of the method, why cannot we accumulate the optimization procedure into the program so as the optimization process can automatically respond to what you want it to do? We aim to incorporate a software product that will optimize the management of the techniques and to achieve that, we have to define the parameters that we are trying to optimize.[42] As for it, you might even think that the study on the bioindustrialization of robotics, the usage of which may serve as a way of automating patching, upgrading, and portability of applications, maybe expanded further. There are so many long-standing problems, ranging from autonomous computing[43] and self-adapting[44] that it will enable us to synthesize with new tools to solve these concerns.

3.4.1.5   ECO-FRIENDLY AI SOFTWARE EXPANSION AND DISPOSITION

It is not feasible to get an AI methodology applied out of an established SE framework and usage case. We need to adjust tactics and goods to the AI-driven SE Age, where people implement AI techniques in many fields. AI algorithms are now operating in industries and worldwide to help build, finance, and validate applications and provide us research, growth, and testing. These modern technologies allow us to make things quicker and more effective and enable current software development techniques and procedures to be utilized for their program.

With the rise in AI, the need to rethink the suitable approaches to involve AI in the app creation phase will fall more and more into place. Let us assume you want to create a vacuum cleaner that operates itself, and you may need to consider an automatic release policy. Maybe automatic patches may function the same way as the unique patch. Still, they might not be as wholly trusted as original patches that they may be used along with the original patches for continuing reversion challenging purposes. If the program is being implemented, it is supposed to take advantage of complex optimization in situ. Under those situations, we would choose to incorporate digital products that will adjust depending on the user's performance and how the program looks like; it would be helpful. The system would require a way to express consciously that they have become disappointed and unhappy with the system because they are indirectly, just by utilizing it, deprived of intrusive use. The device would be futile to query the user smartly every few seconds if they are pleased; the framework must be configured to actively display this before control the replacement of nonfunctional possessions that it tries to simplify.

3.4.1.6   IMPERFECT CLEARNESS

The philosophy of SE is founded around turning large structures into more straightforward, more functional blocks as is conceivable, as ideal to cluster identical chunks into separate blocks at diverse stages of thought which have a comparable significance. While AI technology algorithms theoretically can allow this calculation automatically for each human, it is not simple to recognize closely in what way it is done or forecast the concept layers that will underlie every such learned method. Furthermore, it is challenging to define a particular functional region, and there are no good tests to explain the principles in the model.[45]

By design, swapping a segmented and comparatively easy pipeline of how the medication is developed has provided substantial advancements within the arenas of "computer visualization and language appreciation".[46] A concern is that precision is being traded for accountability. This is an unavoidable condition. There is a lot of specificity in the explanation of the algorithm. If the dilemma is fully known by intuition alone, it would not be complicated at all. The more complicated their simulations are, the more impossible they to replicate. Thus, because some elements of their model are fundamental, the model would encompass as much knowledge as their model.[47]

3.4.1.7   TROUBLESHOOTING

Further, we do not entirely grasp NN's internal mechanisms, which mark it impossible to repair them historically. Think about a neural network as a mixture of functional modules and memories, as well as a sharing of memory through several computers.[48] We may claim that the libraries, such as Tensor Flow, for both the Deep Down and Hyper Face applications are a little outdated, and programming in the language could be much more complicated when opposed to regression models. This pattern is distinct from modern-day Distributed SE Tools. All marks share that the creator lazily executes the design and only continues to manually execute the code in a procedural order.[49]

Supplementary agendas, such as "PyTorch," do not have the sluggish graphical implementation issue. They also do not have a significant amount of adoption for DL within the AI culture. In addition, if it is feasible to find a way to set breakpoints in the source code and pass through it, it is just not achievable in reality.[50] That will be a full inspection of the millions of criteria. Compared, bugs in the code line of a program will distort, cannot be vigorously found during the phase of either compilation or runtime. The knowledge helps the developers or data scientists to accomplish their training targets, such as estimating a global error estimation. After a fair deal of mental processing, the system's power requires a lot of time to build up, and this assures there is little way of ensuring that the system achieves a certain degree of output. They need extensive investments in engineering to make.[51]

3.4.1.8 TESTING

AI software needs comprehensive research to build the statistical pipeline, the training models, and the development facilities. Since AI systems operate with a high volume of data, it is crucial to verify their data. While there are just a few tools for processing data, several other agencies are running behavior on data. When data is checked on a tiny sample, the trend is to see fluctuation, but it is hard to construct a rigorous examination. It is impossible to have a collection of choices that address all of the edge cases in the entire dataset. Also, since the outer universe is complex and evolves with time, new edge cases can have to be introduced to the method later.

It is, by necessity, hard to debug the device because of the nondeterministic existence of the training algorithms. Another problem with the current research might be that the data and model used in standard mode may vary substantially in the deployment process from the testing or training process. This could contribute to skew in the model efficiency standard.[52] When the experiments are in progress, so the findings would be vital.

3.4.2 MANUFACTURE CHALLENGES

It is imperative to yield the benefit of modern hardware when aimed at DL. This presents a significant obstacle in ensuring regular informs to advanced applications besides handling related dependences. Careful and intelligent monitoring is needed to identify the problems posed by altering behavior independences, counting databases changed. If the models were used straight in manufacturing by the utilities, they might change the fact they are attempting to clarify. (This is not possible, but it would not be crazily complicated to arrange.) In the absence of an adequate amount of input data, the model learns by utilizing the minimal amount of online input to change the model's parameters during preparation.[52]

3.4.2.1   ADDICTION MANAGEMENT

There are conventional SE assumptions that say that hardware is at its plumpest when it is a nonissue in the networks, and even if the hardware is an actual issue, it is the worst. AI schemes are mainly skilled on GPUs, making a "40–100× speedup" over traditional CPUs. The efficiency of the graphics card has increased dramatically with the development of the latest technologies. Several new GPUs are launched annually. As each phase substantially improves efficiency[53] and hardware presentation converts straight into decreased preparation period (or improved outcomes for equal preparation period), there is a clear motivation for software to track hardware progress closely.

AI app systems are continually revised weekly and often daily, and upgrades usually result in significant changes. Small and so-called proof-of-concept pumps that are ideal for testing purposes are less workable for natural production systems due to high wear, large seals, and friction, among others. DL functions even as the amount of functionality rises as much as when the number of examples increases. As there is a broad spectrum from 2 h to over a week for a good completion, there is a great incentive to improve output using the new hardware and applications. Changing design, software, and functionality of the devices used to test cognitive performance not only results in losing the opportunity to sustain reproducible outcomes but also would possibly result in considerable expenditure associated with maintaining the software and hardware updated.

3.4.2.2   OBSERVING AND SORTING

First, to depict an AI system, as an illustration, is easy. The representation is adequate as an approximation, but when you begin to concentrate on creating a production-ready AI system, it becomes evident that this is a far more complicated and challenging job. In real-world AI implementations outside toy scenarios, it may become impossible to shelter entire border cases that may exist after the concept has been implemented in development. Since several people fail to understand the work required to support the AI framework installed over time, it may often be said that AI systems need a large amount of management and maintenance.[54]

The AI framework can be retrained as the configuration and organization of the users' lives shift. When the external environment around us turns, our AI reacts in a manner that no longer takes the course we wish it to without needing to take any initiative to make it do so. In this case, unit tests are functional because they avoid the issues at hand from extending to the remainder of the method. Senior levels that have been calculated by humans on a scale from zero to two are no longer accurate due to drifts of data obtained in the outside world. The tracking of device output can be beneficial. Nevertheless, it will be an intelligent choice to select which indicators to be tracked.[55]

3.4.2.3   UNINTENTIONAL RESPONSE LOOPS

Because the AI framework is open ended, it is the case that external data often power it. Despite some carefully developed and implemented models, their final output would still depend on the environment's data (context) generated (outside the model). Furthermore, in particular, in replicas implemented in a broad statistics setting (like ML schemes frequently do), there is a possibility of generating an unwanted response eye where real-world procedures respond to the typical somewhat additional method around. Visualize using a forecasting tool to estimate potential residential real estate values.[55] When one predictive system has become exceedingly common, its concept will become a self-fulfilling prophecy. If the temperature is modified, the temperature-dependent rate of exchange becomes positive, which is unpredictable and under the influence of positive feedback.[56]

3.5   CONCLUSION

One firm conclusion is that while AI technology has produced encouraging outcomes, it is still great essential for more investigation and expansion on creating high-quality, ready-to-use AI systems quickly and efficiently. Traditional SE includes suitable standard methods and techniques for installing, writing checks, and debugging codes. However, studying DL algorithms and other similar algorithms is hard to come by. Along with the AI community, the SE group wants to do a study of such detrimental consequences and work together to resolve them. By addressing the above concerns, AI technology should be rendered open not just to academics and big corporations, nonetheless similarly to the overwhelming bulk corporations across the domain.

We have progressed from tiny, localized, enclosed, healthy distinct constructions to large-scale creation and management of linked, intelligent, dynamic, interconnected structures for many decades. The engineering essence of the challenges we look at as SE, like deafening, partly and loosely defined technology domains with numerous overlapping, overlapping, and evolving goals, is pulling us from a romantic paradise of flawless architecture to an additional practical, yet incomplete, the field of engineering optimization. The shift in the program's design causes us to amend the production and implementation strategies. Not at all astonishment that AI strategies are well adapted toward this evolving environment because AI influences them; the model of a chaotic, ill-defined, overlapping, opposing, linked, dynamic, interconnected framework.

KEYWORDS

- **artificial intelligence**
- **software engineering challenges**
- **smart optimization**

REFERENCES

1. Krizhevsky, A.; Sutskever, I.; Hinton, G. E. Imagenet Classification with Deep Convolutional Neural Networks. *Adv. In. Neu. Inf. Proc. Syst.* **2012,** 1097–1105.
2. Radford, A.; Metz, L.; Chintala, S. Unsupervised Representation Learning with Deep Convolutional Generative Adversarial Networks. *ICLR,* **2016,** 2–8.
3. Mikolov, T.; Karafi´at, M.; Burget, L.; Cernock`y, J.; Khudanpur, S. Recurrent Neural Network Based Language Model. *Inter. Sp.* **2010,** 3–10.
4. Mnih, V.; Kavukcuoglu, K.; Silver, D.; Rusu, A. A.; Veness, J.; Bellemare, M. G.; Graves, A.; Riedmiller, M.; Fidjeland, A. K.; Ostrovski, G. Human-Level Control through Deep Reinforcement Learning. *Nature* **2015,** 529–533.

5. Bengio, Y.; Bengio, S. Modeling High-Dimensional Discrete Data with Multi-Layer Neural Networks. *Adv. Neu. Infor.Proc. Syst*, **2000**, 400–406.

6. Boureau, Y. I.; Cun, Y. L. Sparse Feature Learning for Deep Belief Networks. *Adv. Neu. Info. Proc. Syst.* **2008**, 1185–1192.

7. Hinton, G.; Deng, L.; Yu, D.; Dahl, G. E.; Mohamed, A. R.; Jaitly, N.; Senior, A.; Vanhoucke, V.; Nguyen, P.; Sainath, T. N. Deep Neural Networks for Acoustic Modeling in Speech Recognition: The Shared Views of Four Research Groups. *IEEE. Sig. Proc. Mag.* **2012**, 82–97.

8. Le, Q. V. Acoustics, Speech and Signal Processing (ICASSP). In *2013 IEEE International Conference on*; IEEE, 2013.

9. Poulding, S. M.; Clark, J. A. Efficient Software Verification: Statistical Testing Using Automated Search. *IEEE Trans. Soft. Eng.* **2010**, 763–777.

10. Staunton, J.; Clark, J. A. 13th Annual Genetic and Evolutionary Computation Conference (GECCO), Dublin, Ireland, July 12–16, 2011.

11. Fatiregun, D.; Harman, M.; Hierons, R. *Source Code Analysis and Manipulation (SCAM 04)*; IEEE Computer Society Press: Los Alamitos, California, USA, September 10–12, 2004.

12. Ryan, C. *Automatic Re-Engineering of Software Using Genetic Programming*; Kluwer Academic Publishers, 2000.

13. Greer, D.; Ruhe, G. Software Release Planning: An Evolutionary and Iterative Approach. *Info. Softw. Technol.* **2004**, 243–253.

14. Mitchell, B. S.; Traverso, M.; Mancoridis, S. *Proceedings of the Working Conference on Software Architecture (WICSA '01)*; IEEE Computer Society: Amsterdam, Netherlands, 2001.

15. Mahdavi, K.; Harman, M.; Hierons, R. M. *IEEE International Conference on Software Maintenance*; IEEE Computer Society Press: Los Alamitos, California, USA, September 25-26, 2003.

16. Asadi, F.; Antoniol, G.; Gu´eh´eneuc, Y. *Proceedings of 2nd International Symposium on Search Based Software Engineering (SSBSE 2010)*; IEEE Computer Society Press: Benevento, Italy, 2010.

17. Yoo, S.; Harman, M.; Ur, S. *3rd International Symposium on Search Based Software Engineering (SSBSE 2011)*; September 10–12, 2011.

18. Linderman, M. D.; Collins, J. D.; Wang, H.; Meng, T. H. *13th International Conference on Architectural Support for Programming Languages and Operating Systems (ASPLOS)*; ACM: Seattle, WA, USA, Mar 15–16, 2008.

19. Buttari, A.; Dongarra, J.; Kurzak, J.; Langou, J,; Luszczek, P.; Tomov, S. *8th International Workshop Applied Parallel Computing (PARA 2006), Vol. LNCS 4699*; Springer: Umea, Sweden, June 9–10, 2006.

20. Karlsson, J.; Ryan, K. A Cost-Value Approach for Prioritizing Requirements. *IEEE Softw.* **1997**, 67–74.

21. Zhang, Y.; Harman, M.; Finkelstein, A. Comparing the Performance of Metaheuristics for the Analysis of Multi-Stakeholder Tradeoffs in Requirements Optimization. *J. Info. Softw. Technol.* **2011**, 761–773.

22. Barreto, A.; Barros, M.; Werner, C. Staffing a Software Project: A Constraint Satisfaction and Optimization Based Approach. *Comp. Oper. Res.* **2008**, 3073–3085.

23. Antoniol, G.; Di Penta, M.; Harman, M. The Use of Search-Based Optimization Techniques to Schedule and Staff Software Projects: An Approach and an Empirical Study. *Soft.—Pract. Exp.* **2011,** 495–519.

24. Mancoridis, S.; Mitchell, B. S.; Chen, Y. F.; Gansner, E. R. *Proceedings; IEEE International Conference on Software Maintenance*; IEEE Computer Society Press, 1999.

25. Praditwong, K.; Harman, M.; Yao, X. Software Module Clustering as a Multi-Objective Search Problem. *IEEE Trans. Softw. Eng.* **2011,** 264–282.

26. Fraser, G.; Arcuri, A. *11th International Conference on Quality Software (QSIC)*; IEEE Computer Society: Madrid, Spain, July 1–2, 2011.

27. Harman, M.; McMinn, P. A Theoretical and Empirical Study of Search Based Testing: Local, Global and Hybrid Search. *IEEE Trans. Softw. Eng.* **2010,** 226–247.

28. Finkelstein, A.; Harman, M.; Mansouri, A. A Search Based Approach to Fairness Analysis in Requirements Assignments to Aid Negotiation, Mediation and Decision Making. *Req. Eng.* **2009,** 231–245.

29. Saliu, M. O.; Ruhe, G. *European Software Engineering Conference and the ACM SIGSOFT International Symposium on Foundations of Software Engineering (ESEC/FSE)*, ACM, September 2007.

30. Simons, C. L.; Parmee, I. C.; Gwynllyw, R. Interactive, Evolutionary Search in Upstream Object-Oriented Class Design. *IEEE Trans. Softw. Eng.* **2010,** 798–816.

31. Antoniol, G.; Gueorguiev, S.; Harman, M. *ACM Genetic and Evolutionary Computation Conference (GECCO 2009)*, Montreal, Canada, July 8–12, 2009.

32. Harman, M.; Krinke, J.; Ren, J.; Yoo, S. *ACM Genetic and Evolutionary Computation Conference (GECCO 2009)*, Montreal, Canada, July 8–12, 2009.

33. Bouktif, S.; Sahraoui, H.; Antoniol, G. *Proceedings of the 8th Annual Conference on Genetic and Evolutionary Computation*, Vol. 2; ACM Press Seattle, Washington, USA, July 8–12, 2006.

34. Dolado, J. J. On the Problem of the Software Cost Function. *Info. Softw. Technol.* **2001,** 61–72.

35. Krogmann, K.; Kuperberg, M.; Reussner, R. Using Genetic Search for Reverse Engineering of Parametric Behaviour Models for Performance Prediction. *IEEE Trans. Softw. Eng.* **2010,** 865–877.

36. Rodriguez, D.; Ruiz, R.; Riquelme-Santos, J. C.; Harrison, R. *3rd International Symposium on Search Based Software Engineering (SSBSE)*; Springer: Hungary, September 4–5, 2011.

37. Harman, M. *15th International Conference on Program Comprehension (ICPC 07)*; IEEE Computer Society Press: Banff, Canada, 2007.

38. Souza, J.; Maia, C. L.; de Freitas, F. G.; Coutinho, D. P. *2nd International Symposium on Search based Software Engineering (SSBSE 2010)*; IEEE Computer Society Press: Benevento, Italy, 2010.

39. Arcuri, A.; Yao, X. *Proceedings of the IEEE Congress on Evolutionary Computation (CEC '08)*; IEEE Computer Society: Hongkong, China, June 1–6, 2008.

40. Weimer, W.; Nguyen, T. V.; Goues, C. L.; Forrest, S. *International Conference on Software Engineering (ICSE 2009)*; Vancouver, Canada, 2009.

41. Langdon, W. B.; Harman, M. Evolving a CUDA Kernel from an Vidia Template. *IEEE Congr. Evol. Comp.* **2010,** 1–8.

42. White, D. R.; Clark, J. *Genetic and Evolutionary Computation Conference (GECCO 2008)*; ACM Press: Atlanta, USA, July 6–8, 2008.
43. Kephart, J. O.; Chess, D. M.; The Vision of Autonomic Computing. *IEEE Comp.* **2003,** 41–50.
44. Filieri, A.; Ghezzi, C.; Tamburrelli, G. A Formal Approach to Adaptive Software: Continuous Assurance of Non-Functional Requirements. *For. Asp. Comp.* **2012,** 163–186.
45. Bengio, Y. *Proceedings of ICML Workshop on Unsupervised and Transfer Learning,* 2012.
46. LeCun, L.; Bengio, Y.; Hinton, G. Deep Learning. *Nature* **2015,** 436–444.
47. Glorot, X.; Bengio, Y. *Proceedings of the Thirteenth International Conference on Artificial Intelligence and Statistics,* 2010.
48. Abadi, M.; Agarwal, A.; Barham, P.; Brevdo, E.; Chen, Z.; Citro, C.; Corrado, G. S.; Davis, A.; Dean, J.; Devin. M. Tensorflow: Large-Scale Machine Learning on Heterogeneous Distributed Systems. *Nature* **2017,** 402–412.
49. Zaharia, M.; Chowdhury, M.; Das, T.; Dave, A.; Ma, J.; McCauley, M; Franklin, M. J. *Proceedings of the 9th USENIX Conference on Networked Systems Design and Implementation*; USENIX Association, 2012.
50. P. Core Team. PyTorch, 2017. http://pytorch.org/ (accessed Aug 7, 2017).
51. Chen, X. W.; Lin, X. Big Data Deep Learning: Challenges and Perspectives. *IEEE Access* **2014,** 514–525.
52. Tata, S.; Popescul, A.; Najork, M.; Colagrosso, M.; Gibbons, J.; Green, A.; Mah, A.; Smith, M.; Garg, D.; Meyer, C. *Proceedings of the 23rd ACM SIGKDD International Conference on Knowledge Discovery and Data Mining*; ACM, 2017.
53. Geer, D. Taking the Graphics Processor beyond Graphics. *Computer* **2005,** 14–16.
54. Breck, E.; Cai, S.; Nielsen, E.; Salib, M.; Sculley, D. What's Your ML Test Score? A Rubric for ML Production Systems. *Rel. Mach. Lear.* **2016,** 125–136.
55. Sculley, D.; Holt, G.; Golovin, D.; Davydov, E.; Phillips, T.; Ebner, D.; Chaudhary, V.; Young, M.; Crespo, J. F.; Dennison, D. Hidden Technical Debt in Machine Learning Systems. *Adv. Neur. Info. Proc. Syst.* **2015,** 2503–2511.
56. Bottou, L.; Peters, J.; Qui ˜nonero-Candela, J.; Charles, D. X.; Chickering, D. M.; Portugaly, E.; Ray, D.; Simard, P.; Snelson, E. Counterfactual Reasoning and Learning Systems: The Example of Computational Advertising. *J. M. Lear. Res.* **2013,** 3207–3260.

CHAPTER 4

Uncertain Multiobjective COTS Product Selection Problems for Modular Software System and Their Solutions by Genetic Algorithm

ANITA R. TAILOR[1][*] and JAYESH M. DHODIYA[2]

[1]*Department of Mathematics, Navyug Science College, Surat, India*

[2]*Department of Applied Mathematics and Humanities, S. V. National Institute of Technology, Surat*

[*]*Corresponding author. E-mail: anitatailor_185@outlook.com*

ABSTRACT

This chapter focuses on a genetic algorithm (GA)-based approach to find the solution of uncertain multiobjective commercial-off-the-shelf (COTS) product selection problem with triangular and/or trapezoidal numbers in objective functions subject to several pragmatic constraints. In this chapter, α–level set is employed to categorize fuzzy decision for the decision maker (DM) to optimize the various scenario of fuzzy objective functions and a fuzzy technique is used to tackle multiobjective optimization problem wherein the DM must specify different aspiration levels (ALs) according to his/her preferences and different shape parameters in the exponential membership function to show the effect of integration on the effective allocation plans of COTS production selection problems. An existent scenario is proposed to represent the significance of the proposed

Computational Intelligence Applications for Software Engineering Problems. Parma Nand, PhD, Rakesh Nitin, PhD, Arun Prakash Agrawal, PhD & Vishal Jain, PhD (Eds.)

approach with data set from realistic circumstances. This research turned out to be a GA-based approach that provides effective output based on analysis to take decision regarding the situation.

4.1   INTRODUCTION

In today's world, software system development has focused on improving the effectiveness of structured programming, modeling languages, and design models. With the growing prerequisite of advanced software systems, the use of commercial-off-the-shelf (COTS) products has steadily increased because of the nature of COTS as reusable software pieces for software developers to create new software systems. Because COTS products are ready-made, software developers can use them "as it is" and integrate them with easy-to-install and accessible system components.[16]

The use of COTS products in software systems offers a number of benefits such as reducing development costs, development time and effort, and improving target software quality. In spite the advantages of COTS product, performance investigation of the resulting system often becomes complex as restricted knowledge about the COTS products. Thus, the selection of best-fit COTS products is not a straightforward task, and also it is tricky to select best-fit COTS product when it is dealing with multiple criteria decision-making procedure. In addition, it is required to go for the accessibility of COTS products that fulfill the functional requirement such as system cost, system quality, and system delivery time. Therefore, an effectual approach is required to select the best fit COTS product. Furthermore, the COTS selection problem considers the following tasks as availability of alternative COTS components, multiple COTS selection objectives, and the selection of a single COTS component per module.[15] COTS selection product of several optimization models is present in the literature to carry out the various features of quality along with the objective of cost minimization. Several studies in the literature have solved COTS selection optimization problems.[1,3,5,7,12,13,19,28,30,32]

In an optimization model of COTS software product, we may often meet uncertain phenomena such as random phenomena, fuzzy phenomena due to the uncertainty, nondeterministic model parameters, etc. Additionally, COTS products are precious by ambiguity and vagueness with linguistic terms which are given by decision maker (DM). In such conditions, COTS product selection problems become an uncertain COTS product selection problem. In the decision-making problems of the real world, the selection of COTS product problem is more valuable through fuzzy theory. The fuzzy models of COTS product selection problem have been depicted in detail in the literature.[5,6,9,10,14,15,24,32]

In this chapter, the problems of uncertain multiobjective (UMO) COTS product selection for triangular fuzzy numbers and trapezoidal fuzzy numbers are studied. To solve these UMOCOTS product selection problems, GA-based approach is used together with possibility distribution. To find the solution of such problems, the given problem is a transformed into crisp multiobjective (CMO) COTS product selection problem using possibility distribution and then into a single objective nonlinear optimization problem (SONOP) and therefore such problem turn into "NP-hard" problem. To solve such problems, GA is a suitable course of action. Thus, this chapter proposes a GA-based approach to ascertain the efficient solutions (ESs) of the Uncertain Multi-objective COTS (UMOCOTS) product selection problem.

4.2 MULTIOBJECTIVE UNCERTAIN OPTIMIZATION MODEL OF THE COTS PRODUCT SELECTION PROBLEM

In this chapter, the design of the modular software system (MSS) is considered. Depending on the software requirements, the MSS can perform multiple functions called programming. Each program consists of several modules that are executed in sequence. For different programs, some of these modules are common. For each module, a number of alternative COTS products are accessible. In software system, each module has different levels of importance that depending on the access occurrence rate.[5,6,14] The diagrammatic description of the software system is described in Figure 4.1.

4.2.1   UMOCOTS PRODUCT SELECTION MODEL WITH TRIANGULAR NUMBERS

This section discusses the UMOCOTS product selection problem with two objectives as quality and purchasing cost of the software system. To formulate the UMO optimization model with triangular number, some notations and parameters are used defined as follows:[6]

- i and j indicate the index of modules (i=1, 2… m) and COTS products, respectively.
- k and η_i indicate the number of modules and the number of COTS alternatives in i^{th} module (j=1, 2...n_i), respectively.
- w_i and T indicate the number of weight of the ith module and the maximum threshold time on the delivery time of the software system, respectively.

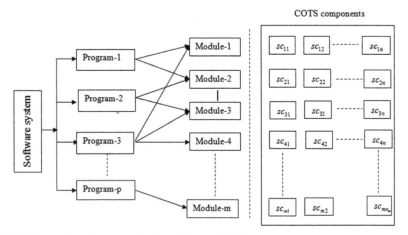

FIGURE 4.1   The diagrammatic description of MSS.

Binary variables:

x_{ij} is symbolize the decision variable as i^{th} COTS alternative is chosen for the ith module or not.

$$x_{ij} = \begin{cases} 1; & \text{if } j^{th} \text{ COTS alternative is chosen for the } i^{th} \text{module} \\ 0; & \text{otherwise} \end{cases}$$

y_1 is symbolize the ith constraints is active or not.

$$y_i = \begin{cases} 0, & \text{if the } i^{th} \text{ constraint is active} \\ 1, & \text{otherwise} \end{cases}$$

Multiobjective COTS selection model

Multiobjective optimization model (MOOM) of the COTS product selection problem with two objectives weighted quality (Z_2) and the total purchasing cost (Z_2) is presented as follows.[6]

MODEL-1:

$$\max Z_1 = \sum_{i=1}^{k} w_i \left(\sum_{j=1}^{\eta_i} q_{ij} x_{ij} \right)$$

$$\min Z_2 = \sum_{i=1}^{k} \sum_{j=1}^{\eta_i} c_{ij} x_{ij}$$

subject to

$$\sum_{j=1}^{\eta_i} d_{ij} x_{ij} \leq T \tag{4.1}$$

$$\sum_{j=1}^{\eta_i} x_{ij} = 1, \, i = 1, 2, ..., m \tag{4.2}$$

$$x_{ij} \in \{0,1\}, \, i = 1, 2, ..., m; \, j = 1, 2, ..., m \tag{4.3}$$

$$x_{rs} - x_{ut_k} \leq My_k; \, k = 1, 2, ..., z \tag{4.4}$$

$$\sum_{k=1}^{z} y_k = z - 1, \tag{4.5}$$

$$y_k \in \{0,1\}, \, k = 1, 2, ..., z \tag{4.6}$$

where q_{ij}, c_{ij}, and d_{ij} indicate the imprecise quality level, the imprecise cost, and the imprecise delivery time of the jth COTS alternative in the ith module, respectively, and eqs. 4.4–4.6 represent the contingent decision constraints, which guarantees that only one out of the z contingent decision constraints is active for any COTS product among two modules if M is very large.

4.2.2   UMOCOTS PRODUCT SELECTION PROBLEM WITH TRAPEZOIDAL NUMBERS

This section discussed the UMOCOTS product selection problem with three objectives cost (Z_1), size (Z_2), and execution time (Z_3). To formulate the uncertain MOOM with trapezoidal number, some notations and parameters are used as follows::[5,14]

Z_1, Z_2, and Z_3 indicate the fuzzy cost c_{ij}, size l_{ij} and execution time t_{ij} of the jth COTS product in the ith module;

p indicates the index of programs $(p = 1, 2,..., P)$.

m, P, and v_p define the number of modules, number of programs, and probability of the use program p, respectively.

d_{ij} and μ_{ij} indicate fuzzy delivery time and probability of failure on demand of the jth COTS product in the ith module, respectively,

S_i and S_p indicate the average number of invocation of ith module and set of modules corresponding to program p, respectively.

R and T indicate minimum and maximum desired levels of the reliability and expected delivery time, respectively.

x_{ij} and y_i symbolize the binary variables.

Multiobjective COTS selection model:

MOOM of COTS product selection problem with three objectives with some constraints are defined as follows:

model-2:

$$\min Z_1 = \sum_{i=1}^{m} \sum_{j=1}^{n_i} c_{ij} x_{ij}$$

$$\min Z_2 = \sum_{i=1}^{m} \sum_{j=1}^{n_i} l_{ij} x_{ij}$$

$$\min Z_3 = \sum_{p=1}^{P} v_p \sum_{i \in S_p} \left(\sum_{j=1}^{n_i} t_{ij} x_{ij} \right)$$

Subject to the constraints:

$$\sum_{j=1}^{n_i} d_{ij}x_{ij} \leq T; i = 1, 2, ..., m \qquad (4.7)$$

$$\prod_{i=1}^{m} \exp\left(-s_i \sum_{j=1}^{n_i} \mu_{ij}x_{ij}\right) \geq R \qquad (4.8)$$

$$\sum_{j=1}^{n_i} x_{ij} = 1, i = 1, 2, ...m \qquad (4.9)$$

$$x_{ij} \in \{0,1\}, i = 1, 2, ..., m; j = 1, 2, ..., m \qquad (4.10)$$

$$x_{rs} - x_{ut_k} \leq My_k; k = 1, 2, ..., z \qquad (4.11)$$

$$\sum_{k=1}^{z} y_k = z - 1, \qquad (4.12)$$

$$y_k \in \{0,1\}, k = 1, 2, ..., z \qquad (4.13)$$

4.3   SOME PRELIMINARIES

4.3.1   POSSIBILISTIC PROGRAMMING APPROACH

The real-world problems with incomplete information are the major issues as it imposes a high level of uncertainty. Even though, past data are presented, the performance of the parameters does not require fulfilling with their past model in the future. To deal with these issues, the uncertain parameters are presented with fuzzy numbers. The theory of possibility is that an important part of the data is subject to human choice and depends on the possibilities in nature. The possibility distribution is calculated from the form of inadequate data and DM's knowledge. Possibilistic programming approach has been used to solve the fuzzy optimization model of numerous significant applications. This approach shifts the fuzzy objective and/or constraints into crisp objective and/or constraints in relation to different scenarios as optimistic, most-likely and pessimistic scenarios. It is also used to maintain the uncertainty of the problem in anticipation of a solution is found.[4,11] As a result, this approach can be used to alter the UMOCOTS selection model into a crisp MOOM.[4,11] Several studies have been used in the literature to address the problem of fuzzy optimization through possibility distribution.[2,4,10,27,31]

4.3.2   TRIANGULAR POSSIBILITY DISTRIBUTION

To represent the triangular uncertain parameters, the triangular possibility distribution is usually applied because of its simplicity and computational effectiveness.[4,11,23,31]

In practical circumstances, DM can create the triangular possibilistic distribution by using the (c_i^o), (c_i^m), and (c_i^p), most optimistic value, the most possible value, and the most pessimistic value, respectively. From Figure 4.2, the cost objective function is defined at three well-known points $\left(c_1^p,0\right)$, $\left(c_1^m,1\right)$, and $\left(c_1^o,0\right)$.

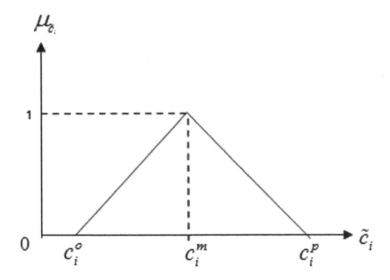

FIGURE 4.2   Triangular possibilistic distribution of c_i.

Source: Reprinted from Ref. [4]. https://creativecommons.org/licenses/by/4.0/

4.3.3 TRAPEZOIDAL POSSIBILITY DISTRIBUTION

To represent the trapezoidal uncertain parameter, the trapezoidal possibility distribution is used.[33]

In particular, for the cost coefficient, $c_i = \left(c_i^o, c_i^{\underline{m}}, c_i^{\overline{m}}, c_i^p\right)$; DM can create the trapezoidal distribution by using $\left(c_i^o\right)$ the most optimistic, $\left(\left[c_i^{\underline{m}}, c_i^{\overline{m}}\right]\right)$ the interval of most likely value that absolutely belongs to the set of available values and $\left(c_i^p\right)$ most pessimistic values. In Figure 4.3, the cost objective function is defined at four well-known points $\left(c_1^p, 0\right), \left(c_1^{\underline{m}}, 1\right), \left(c_1^{\overline{m}}, 1\right)$, and $\left(c_1^o, 0\right)$.

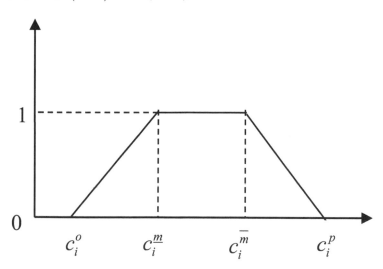

FIGURE 4.3 Trapezoidal possibilistic distribution of C_i.

4.3.4 α-LEVEL SETS

α-Level set is the most fundamental theory to set up the connection between fuzzy and traditional set theory that is initiated by Zadeh in 1965.[29] α-level set is used to fit the confidence of DM in his/her fuzzy decision; it is also called as a confidence level. The small value of α will leads to a large expansion of the objective function value, which indicates that the obtained solution corresponds to a higher level of pessimism and uncertain, whereas the large value of α will give a solution with small dispersions in the objective function value, which indicates that obtained the solution corresponds to a high level of optimism in the DM's fuzzy judgment. There are various studies in the literature that have used the α-level sets concept to find a solution to a problem related to the fuzzy optimization.[4,5,11,20–22,27,31]

4.3.5   ASPIRATION LEVEL

In many UMOCOTS product selection problems, it is imperative to obtain a solution that faithfully reflects the DM's judgment. In order to make decisions based on the diversity of value-judgment and complicated modification in the nature of decision making, the aspiration level (AL) has provided the satisfactory trading method that can work well not simply with the quantitative DM judgment but also with the dynamics of significance opinion of them. AL does not necessitate any uniformity of judgment of DM because DM often changes his/her attitude even throughout the decision-making process.[17,18]

4.3.6   POSITIVE IDEAL SOLUTION (PIS) AND NEGATIVE IDEAL SOLUTION (NIS)

The minimum and the maximum value of an objective function is defined by PIS and NIS, respectively. They are required to compute the membership value for all objective functions.[4,5,31]

4.3.7   EXPONENTIAL MEMBERSHIP FUNCTION

In the UMOCOTS product selection problem, the different ALs of DM are distinguished by fuzzy membership functions for the objective functions. Moreover, the membership function is used for describing the performance of the vague information, utilization of the fuzzy numbers of DM, preferences towards uncertainty, etc. The exponential membership function (EMF) gives healthier demonstration then others and provides the flexibility to express grade of exactness in parameter values. It also reflects reality better than the linear membership function.[4,5,8,27]

If Z_{ij}^{PIS} and Z_{ij}^{NIS} are PIS and NIS of objective Z_{ij}, the EMF $\mu_{Z_{ij}}^{E}$ is defined by

$$
\mu_{Z_{ij}}^{E}(x) = \begin{cases} 1; & \text{if } Z_{ij} \leq Z_{ij}^{PIS} \\ \dfrac{e^{-S\psi_{ij}(x)} - e^{-S}}{1 - e^{-S}}, & \text{if } Z_{ij}^{PIS} < Z_{ij} < Z_{ij}^{NIS} \\ 0; & \text{if } Z_{ij} \geq Z_{ij}^{NIS} \end{cases}
$$

(4.14)

where $\psi_{ij}(x) = \dfrac{Z_{ij} - Z_{ij}^{PIS}}{Z_{ij}^{NIS} - Z_{ij}^{PIS}}$ and $S \neq 0$ are shape parameters (SPs), which are specified by DM such that $\mu_{Z_k}(x) \in [0,1]$. The membership function is strictly convex (concave) for $S < 0$ $(S > 0)$ in $[Z_{ij}^{PIS}, Z_{ij}^{NIS}]$.

4.3.8 GENETIC ALGORITHM

Considering the adaptive evolution and usual selection of biological systems, GA is a distinguished arbitrary search and global optimization method. This is also a suitable method to solve discrete, nonlinear, and nonconvex global optimization problems rather than any traditional method because it seeks the optimal solution by simulating the natural evolutionary process, and is performed by imitating the evaluation principle and chromosome processing work in usual genetics.[4,5,8,20,23,25-27] It has proven many key advantages like strong robustness, convergence to large-scale most favorable, and analogous search competence. GA has high-quality acknowledgment in determining a variety of NP-hard problems.

In GAs, chromosomes are coded along with the problem, and fitness function is used to measure chromosomes. Apply gene operators such as selection, crossover, and mutation to produce the new population.[4,5,8] Thus, the newly generated population included some of the parent population and child population, which can be used to find an ES.

4.3.9   CONVERGENCE CRITERIA

If after getting some particular value, that is, optimum value, we say GA is converged. For an NP-hard problem, GA has converged at a large-scale optimum is impractical, unless you already have the test data set with the best solution known. Moreover, the dimension of the problem so affects the convergence of GA. One can define the size of the chromosome (number of solutions) based on the problem's parameter. Also, one should note that increasing the size of the chromosome affects the GA's rate of convergence.[4,5]

4.4   FORMULATION OF UMOCOTS OPTIMIZATION MODELS USING POSSIBILITY DISTRIBUTION

4.4.1   UMOCOTS OPTIMIZATION MODEL-1

To convert the model-1 into crisp MOOM, the triangular possibility distribution strategy is taken.[4,11,27,32]

So, to represent the three distinct scenarios with α-set conception, the UMOCOTS product selection problem of model-1 is changed into CMOCOTS product selection problems, which is as follow:

Model-1.1:

$$\left(\max Z_{11}, \max Z_{12}, \max Z_{13}, \min Z_{21}, \min Z_{22}, \min Z_{23}\right) =$$

$$\left(\begin{array}{c} \sum_{i=1}^{k} w_i \left(\sum_{j=1}^{n_i} (q_{ij})^o_{\alpha} x_{ij}\right), \sum_{i=1}^{k} w_i \left(\sum_{j=1}^{n_i} (q_{ij})^m_{\alpha} x_{ij}\right), \sum_{i=1}^{k} w_i \left(\sum_{j=1}^{n_i} (q_{ij})^p_{\alpha} x_{ij}\right), \\ \sum_{i=1}^{n} \sum_{j=1}^{n} (c_{ij})^o_{\alpha} x_{ij}, \sum_{i=1}^{n} \sum_{j=1}^{n} (c_{ij})^m_{\alpha} x_{ij}, \sum_{i=1}^{n} \sum_{j=1}^{n} (c_{ij})^p_{\alpha} x_{ij} \end{array}\right)$$

Subject to the constraints: (4.1)–(4.6).

4.4.2   UMOCOTS OPTIMIZATION MODEL-2

To convert the model-2 into auxiliary MOOM, the trapezoidal possibility distribution strategy is taken. So, to represent the four distinct scenarios with α-set conception, the UMOCOTS product selection problem of model-2 is changed into CMOCOTS product selection problems, which is as follow:

model-2.1:

$$
\begin{pmatrix} \min Z_{11}, \min Z_{12}, \min Z_{13}, \min Z_{14}, \min Z_{21}, \min Z_{22}, \min Z_{23}, \min Z_{24}, \\ \min Z_{31}, \min Z_{32}, \min Z_{33}, \min Z_{34} \end{pmatrix} =
$$

$$
\begin{pmatrix}
\sum_{i=1}^{m}\sum_{j=1}^{n_i}\left(c_{ij}\right)_{\alpha}^{o} x_{ij},\ \sum_{i=1}^{m}\sum_{j=1}^{n_i}\left(c_{ij}\right)_{\alpha}^{m} x_{ij},\ \sum_{i=1}^{m}\sum_{j=1}^{n_i}\left(c_{ij}\right)_{\alpha}^{\overline{m}} x_{ij},\ \sum_{i=1}^{m}\sum_{j=1}^{n_i}\left(c_{ij}\right)_{\alpha}^{p} x_{ij}, \\
\sum_{i=1}^{m}\sum_{j=1}^{n_i}\left(l_{ij}\right)_{\alpha}^{o} x_{ij},\ \sum_{i=1}^{m}\sum_{j=1}^{n_i}\left(l_{ij}\right)_{\alpha}^{m} x_{ij},\ \sum_{i=1}^{m}\sum_{j=1}^{n_i}\left(l_{ij}\right)_{\alpha}^{\overline{m}} x_{ij},\ \sum_{i=1}^{m}\sum_{j=1}^{n_i}\left(l_{ij}\right)_{\alpha}^{p} x_{ij}, \\
\sum_{p=1}^{P}v_p\sum_{i\in S_p}\left(\sum_{j=1}^{n_i}\left(t_{ij}\right)_{\alpha}^{o} x_{ij}\right),\ \sum_{p=1}^{P}v_p\sum_{i\in S_p}\left(\sum_{j=1}^{n_i}\left(t_{ij}\right)_{\alpha}^{m} x_{ij}\right),\ \sum_{p=1}^{P}v_p\sum_{i\in S_p}\left(\sum_{j=1}^{n_i}\left(t_{ij}\right)_{\alpha}^{\overline{m}} x_{ij}\right),\ \sum_{p=1}^{P}v_p\sum_{i\in S_p}\left(\sum_{j=1}^{n_i}\left(t_{ij}\right)_{\alpha}^{p} x_{ij}\right)
\end{pmatrix}
$$

Subject to the constraints: (4.7)–(4.13).

4.5   SOLUTION APPROACH FOR SOLVING UNCERTAIN MOOMS OF COTS SELECTION PROBLEM WITH POSSIBILITY DISTRIBUTION

This section introduces a GA-based approach for the UMOCOTS product selection problem to determine the finest ESs. As well as this approach provides better litheness to solve such problem in terms of different variety of ALs for all objective functions.

4.5.1   STEPS FOR FINDING THE SOLUTION OF UMO OPTIMIZATION MODELS OF COTS SELECTION PROBLEM, USING GA-BASED APPROACH

The step by step description of a GA-based approach to find ESs to the UMOCOTS product selection problem(s) are as follows:

Step-1: Formulate the model-1 and/or model-2 of multiobjective COTS selection problem(s), using suitable triangular and/or trapezoidal possibility distribution.

Step-2: Define the model-1.1 and/or model-2.1 along with α.

Step-3: Convert the maximum objective function of the problem into minimum form of model-1.1 and/or model-2.1.

Step-4: Find out PIS and NIS for all Z_{ij} of the model-1.1 and/or model-2.1.

Step-5: Determine $m_{Z_{ij}}^E$ for each objective function.

Step-6: In this step, MOOMs of the COTS selection problem of model-1.1 and/or 2.1 converted in the SONOPs that are as follows.

(model-1.1.1) $\max W = \prod\limits_{i=1}^{2}\prod\limits_{j=1}^{3} \mu_{Z_{ij}}$

Subject to the constraints: (4.1)–(4.6) and

$$\mu_{Z_{ij}}(x) - \overline{\mu_{Z_{ij}}}(x) \geq 0; \ i = 1,2, \ j = 1,2,3 \tag{4.15}$$

and/or **(model-2.1.1)** $\max W = \prod\limits_{i=1}^{3}\prod\limits_{j=1}^{4} \mu_{Z_{ij}}$

Subject to the constraints (4.7)–(4.13) and

$$\mu_{Z_{ij}}(x) - \overline{\mu_{Z_{ij}}}(x) \geq 0; \ i = 1,2,3, \ j = 1,2,3,4 \tag{4.16}$$

Step-7: To find different allocations plans for model-1.1.1 and/or model-2.1.1, which is developed in step-6 using GA with verities of the SPs.

  (I) **Encoding of Chromosomes:**
        Encoding of chromosomes is the primary task to solve any problem with GA. To create the solution for the SONOP through GA, the data structure of chromosomes in the encoding space must be considered, which indicates the solution to the problem. In the data structure of chromosomes, we set 0's to all $N \times M$ gene of a chromosomes and thereafter arbitrarily select gene of the chromosome, put 1's in every rows and columns of chromosomes precisely one time such that it satisfies constraints (4.1)–(4.6) and/or (4.7)–(4.13) of single optimization problem model. Each component in the string can be stated as 2^r; $r = 0 : n-1$.

 (II) **Evaluate the fitness function:**
        In GA, the fitness function is a key factor to solve any optimization problem. With the satisfaction of constraints (4.1)–(4.6), (4.15), and/or (4.7)–(4.13), (4.16), the objective function of

model-1.1.1 and/or model-2.1.1 is evaluated to check the fitness of chromosomes.

(III) Selection:

The Selection operator is utilized to decide how chromosomes are preferred from the present population and will treat as parents in favor of the next genetic operations, which will have the highest fitness. To find the solution of UMOCOTS product selection problem(s), tournament selection is used by reason of its competence and simple execution. In this selection, n-chromosomes are arbitrarily preferred from the population and judged against each other. The maximum fit chromosome (winner) is preferred for the subsequent generation, whereas others are ineligible. This process is unremitting until the population size and the number of winners become same.[4,5,8,20,23,25–27]

(IV) Crossover:

Crossover is a course of action of obtaining two-parent solutions and producing a child from them. For the UMOCOTS product selection problem(s), two-point crossover is used to breed new individuals. In this crossover, swap the gene values between the randomly two crossover points in two selected parent chromosomes to breed the new individuals.[4,5,7,25–27]

(V) Construct the threshold:

In order to preserve the population's diversity later than the crossover, a threshold is built to produce UMOCOTS product selection problem(s)'s solution. In such step, select some chromosomes from the parent and child population sets in favor of the new iterations. To make the threshold, first, the whole population must be sorting in an ascending/descending order according to its objective function values and then select the encoded entity strings from all groups. Rest on these objective function values, the population is separated into four groups: above $\mu + 3 \times \sigma$, among $\mu + 3 \times \sigma$ and μ, among μ and $\mu - 3 \times \sigma$, and below $\mu - 3 \times \sigma$. As a result, the capable string cannot be missed, where σ and μ are standard deviation and expected value of the objective function values of parenthood and childhood population, respectively.[4,5,25–27]

(VI) Mutation:

Here, we have used a swap mutation operator among the available numerous mutation operators to improve the missing genetic materials. It is also used to arbitrarily allocate genetic information. In this mutation, two arbitrary spots are chosen in a string and the consequent values are swapped among spots.[4,5,25–27]

(VII) Termination criteria:

A GA is run over a given number of iterations until a termination condition has been reached, that is, the population's best fitness has not changed for a convinced number of generations.

As a result, two cases are executed to find the SONOP's solution using GA: (1) Without mutation and (2) With mutation. In both cases, an algorithm converged to an optimal solution of SONOP, and at the end ES of UMOCOTS product selection problem(s) for MSS with different combinations of SP and AL are obtained.

If DM approved the achieved solution, considers it an ideal solution, and terminate the solution procedure otherwise change the SP, AL, and reiterate the step-2 to 7 till an agreeable ES is achieved for model-1.1 and/or model-2.1.

4.6 UMOCOTS PRODUCT SELECTION PROBLEMS

To express the applicability of the UMOCOTS product selection problems, an industrial case scenario has been adopted from the article.[6] COTS product selection is utilized to build up a software system as an enterprise resource planning (ERP) software system on behalf of small- and medium-size enterprises and also COTS is referred to as an application software package (ASP). This system is an assembly of ASPs that are compiled by a set of standard functional requirements. ASPs are differentiated by a retailer to provide a set of standard functions that can be customized to suit the specific needs of each customer. The diagrammatic analysis of the ERP programming structure is given in Figure 4.4.

There are three functional requirements as Finance (p_1), Operations (p_2), and Marketing (p_3) of the ERP system. There are four modules as accounts (m_1), inventory (m_2), sales order (m_3), and sales promotion (m_4), which are defined by the software development team of the company. p_1 calls two modules as m_1 and m_2, p_2 calls two modules as m_2 and m_3, and p_3 calls three modules as m_1, m_3, and m_4. Module-1 consist of three COST alternatives as $sc_{11}, sc_{12}, sc_{13}$, module-2 consists of two COTS alternative as

sc_{21}, sc_{22}, module-3 consists of three COTS alternative as sc_{31}, sc_{32}, sc_{13}, and module-4 consist of three COTS alternative as sc_{41}, sc_{42}, sc_{43}, respectively.

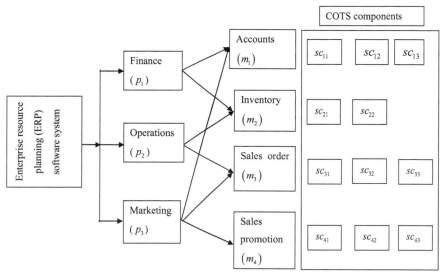

FIGURE 4.4 The diagrammatic structure of the ERP software.

Source: Adapted from Refs. [5, 14, 15].

4.6.1 PROBLEM-1: UMOCOTS PRODUCT SELECTION PROBLEM WITH TRIANGULAR FUZZY NUMBERS

To validate the suitability of the GA-based approach on COTS selection model-1, the UMOCOTS product selection problem has been considered by Gupta et al.[6] The different parameters of COTS selection problem with triangular fuzzy numbers are given in Table 4.1.

TABLE 4.1 The Different Parameters of COTS.

Quality			
$q_{11} = (0.8, 0.83, 0.90)$	$q_{21} = (0.83, 0.85, 0.88)$	$q_{11} = (0.82, 0.85, 0.90)$	$q_{11} = (0.84, 0.90, 0.95)$
$q_{12} = (0.78, 0.82, 0.87)$	$q_{22} = (0.82, 0.88, 0.93)$	$q_{12} = (0.77, 0.79, 0.84)$	$q_{12} = (0.81, 0.88, 0.96)$
$q_{13} = (0.74, 0.78, 0.82)$		$q_{13} = (0.85, 0.90, 0.98)$	$q_{13} = (0.77, 0.81, 0.87)$

TABLE 4.1   The Differe

Cost			
$c_{11} = (5,10,13)$ $c_{12} = (7,9,14)$ $c_{13} = (5,8,11)$	$c_{21} = (1,8,12)$ $c_{22} = (3,9,11)$	$c_{31} = (7,8,14)$ $c_{32} = (4,7,11)$ $c_{33} = (6,9,13)$	$c_{41} = (3,9,11)$ $c_{42} = (2,8,12)$ $c_{43} = (6,8,11)$
Delivery time			
$d_{11} = (4,5,7)$ $d_{12} = (3,4,6)$ $d_{13} = (5,7,10)$	$d_{21} = (5,7,10)$ $d_{22} = (3,6,8)$	$d_{31} = (1,3,6)$ $d_{32} = (3,7,9)$ $d_{33} = (3,5,8)$	$d_{41} = (3,6,8)$ $d_{42} = (4,8,10)$ $d_{43} = (3,6,8)$
Weight			
0.415	0.268	0.177	0.140
Maximum desired level			
$T = (3,8,12)$			
Contingent decision constraints			
$x_{42} \le x_{11}$ or $x_{42} \le x_{13}$			

From Table 4.1, the MOCOTS selection problem model-1.1 is written as follows:

$$\max Z_1 = 0.415 \times \big((0.8,0.83,0.9)x_{11} + (0.78,0.82,0.87)x_{12} + (0.74,0.78,0.82)x_{13} \big) +$$
$$0.268 \times \big((0.83,0.85,0.88)x_{21} + (0.82,0.88,0.93)x_{22} \big) +$$
$$0.177 \times \big((0.82,0.85,0.9)x_{31} + (0.77,0.79,0.84)x_{32} + (0.85,0.90,0.98)x_{33} \big) +$$
$$0.14 \times \big((0.84,0.90,0.95)x_{41} + (0.81,0.88,0.96)x_{42} + (0.77,0.81,0.87)x_{43} \big)$$

$$\min Z_2 = (5,10,3)x_{11} + (7,9,14)x_{12} + (5,8,11)x_{13} + (1,8,12)x_{21} + (3,9,11)x_{22} +$$
$$(7,8,14)x_{31} + (4,7,11)x_{32} + (6,9,13)x_{33} + (3,9,11)x_{41} + (2,8,12)x_{42} + (6,8,11)x_{43}$$

subject to the constraints:

$$(5,7,10)x_{21} + (3,6,8)x_{22} \le (3,8,12), (5,7,10)x_{21} + (3,6,8)x_{22} \le (3,8,12),$$
$$(1,3,6)x_{31} + (3,7,9)x_{32} + (3,5,8)x_{33} \le (3,8,12), (3,6,8)x_{41} + (4,8,10)x_{42} + (3,6,8)x_{43} \le (3,8,12),$$

$$x_{11} + x_{12} + x_{13} = 1,$$

$$x_{21} + x_{22} = 1,$$

$$x_{31} + x_{32} + x_{33} = 1,$$

$$x_{41} + x_{42} + x_{43} = 1,$$

$$x_{21} - x_{11} \leq 5y_1,$$

$$x_{42} - x_{13} \leq 5y_2,$$

$$y_1 + y_2 = 1,$$

$$y_1, y_2 \in \{0,1\},$$

$$x_{ij} \in \{0,1\}, \forall i, j.$$

To evaluate the above problem, the model is coded. Table 4.2 presents the PIS and NIS for Z_{1j} and $Z_{2j}; j = 1,2,3$ of model-1.1 at $\alpha = 01$, 0.5 and 0.9. To maintain the uniformity, maximize quality function is converted into minimized form.

Different combinations of SP and AL are shown in Table 4.3, which is given by DM.

TABLE 4.2 PIS and NIS for Z_{1j} and $Z_{2j}; j = 1,2,3$ at a = 0.1, 0.5 and 0.9.

z_{12} level	Solutions	Objectives					
		z_{12}	z_{12}	z_{22}	z_{22}	z_{22}	z_{23}
$\alpha = 0.1$	PIS	0.516477	0.51636	0.57116	18.7	33	45.6
	NIS	0.541248	0.54843	0.60675	24.1	36	48.5
$\alpha = 0.5$	PIS	0.5102	0.51221	0.538505	25	33	40
	NIS	0.54444	0.54843	0.58083	28.5	37	42.5
$\alpha = 0.9$	PIS	0.5118	0.5122	0.517469	29.5	31	32.4
	NIS	0.5712	0.5731	0.580501	35	37	38.1

TABLE 4.3 Different Combinations of SPs and ALs for Problem-1.

Case	Shape parameters	Aspiration levels
1	(−5, −5)	0.8, 0.9
2	(−5, −1)	0.75, 0.7
3	(−5, −2)	0.8, 0.7

TABLE 4.3   *(Continued)*

Case	Shape parameters	Aspiration levels
4	$(-3, -5)$	0.6, 0.8
5	$(-2, -3)$	0.5, 0.8
6	$-1, -2)$	0.4, 0.65

Table 4.4 describes the computational results and its corresponding optimal allocations according to distinct combinations, which are shown in Table 4.3.

TABLE 4.4   Summary Results of Problem-1 at $\alpha = 0.1, 0.5, 0.9$.

α	Case	Degree of satisfaction	Membership values $\left(\mu_{Z_{1j}}, \mu_{Z_{2j}}, \mu_{Z_{3j}}\right)$	Objective values Z_1, Z_2, Z_3	Optimum allocations
$\alpha = 0.1$	1	0.8021	(0.8752, 0.8656, 0.8021) (1, 0.9709, 0.9987)	(0.8018, 0.8420, 0.8870) (18.7, 34, 45.7)	$sc_{12}, sc_{22}, sc_{32}, sc_{41}$
	2	0.7698	(0.8752, 0.8656, 0.8021) (1, 0.7698, 0.9769)	(0.8018, 0.8420, 0.8870) (18.7, 34, 45.7)	$sc_{12}, sc_{22}, sc_{32}, sc_{41}$
	3	0.8021	(0.8752, 0.8656, 0.8021) (1, 0.8517, 0.9888)	(0.8018, 0.8420, 0.8870) (18.7, 34, 45.7)	$sc_{12}, sc_{22}, sc_{32}, sc_{41}$
	4	0.6478	(0.7419, 0.7286, 0.6478) (1, 0.9709, 0.9987)	(0.8018, 0.8420, 0.8870) (18.7, 34, 45.7)	$sc_{12}, sc_{22}, sc_{32}, sc_{41}$
	5	0.5450	(0.6440, 0.6294, 0.5450) (1, 0.9100, 0.9943)	(0.8018, 0.8420, 0.8870) (18.7, 34, 45.7)	$sc_{12}, sc_{22}, sc_{32}, sc_{41}$
	6	0.5140	(0.5289, 0.5140, 0.4361) (1, 0.8517, 0.9888)	(0.8018, 0.8420, 0.8870) (18.7, 34, 45.7)	$sc_{12}, sc_{22}, sc_{32}, sc_{41}$

$\alpha = 0.5$	1	0.9071	(0.9277, 0.9071, 0.9142) (1, 0.9241, 0.9883)	(0.8259, 0.8461, 0.8753) (25, 35, 40.5)	$sc_{11}, sc_{22}, sc_{32}, sc_{41}$
	2	0.7599	(0.8106, 0.8300, 0.7599) (0.9106, 0.8347, 0.8711)	(0.8197, 0.8420, 0.8670) (25.5, 34, 40.5)	$sc_{12}, sc_{22}, sc_{32}, sc_{41}$
	3	0.7311	(0.9277, 0.9071, 0.9142) (1, 0.7311, 0.9230)	(0.8259, 0.8461, 0.8753) (25, 35, 40.5)	$sc_{11}, sc_{22}, sc_{32}, sc_{41}$
	4	0.7896	(0.8238, 0.7896, 0.8010) (1, 0.9241, 0.9883)	(0.8259, 0.8461, 0.8753) (25, 35, 40.5)	$sc_{11}, sc_{22}, sc_{32}, sc_{41}$
	5	0.6979	(0.7386, 0.6979, 0.7113) (1, 0.8176, 0.9569)	(0.8259, 0.8461, 0.8753) (25, 35, 40.5)	$sc_{11}, sc_{22}, sc_{32}, sc_{41}$
	6	0.5857	(0.6310, 0.5857, 0.6004) (1, 0.7311, 0.9230)	(0.8259, 0.8461, 0.8753) (25, 35, 40.5)	$sc_{11}, sc_{22}, sc_{32}, sc_{41}$
$\alpha = 0.9$	1	0.9096	(0.9185, 0.9155, 0.9096) (0.9750, 0.9709, 0.9640)	(0.8305, 0.8339, 0.8384) (31.2, 33, 34.5)	$sc_{12}, sc_{21}, sc_{32}, sc_{41}$
	2	0.7408	(0.9185, 0.9155, 0.9096) (0.7892, 0.7698, 0.7408)	(0.8305, 0.8339, 0.8384) (31.2, 33, 34.5)	$sc_{12}, sc_{21}, sc_{32}, sc_{41}$
	3	0.8295	(0.9185, 0.9155, 0.9096) (0.8661, 0.8517, 0.8295)	(0.8305, 0.8339, 0.8384) (31.2, 33, 34.5)	$sc_{12}, sc_{21}, sc_{32}, sc_{41}$
	4	0.8841	(0.9083, 0.9047, 0.8963) (0.9121, 0.9241, 0.8841)	(0.8305, 0.8339, 0.8384) (31.2, 33, 34.5)	$sc_{12}, sc_{21}, sc_{32}, sc_{41}$
	5	0.8107	(0.8111, 0.8165, 0.8107) (0.8111, 0.8176, 0.8113)	(0.8375, 0.8420, 0.8470) (32.3, 34, 35.3)	$sc_{12}, sc_{22}, sc_{32}, sc_{41}$
	6	0.6583	(0.7665, 0.7598, 0.7446) (0.7072, 0.7311, 0.6583)	(0.8410, 0.8446, 0.8490) (32.4, 34, 35.7)	$sc_{12}, sc_{21}, sc_{31}, sc_{41}$

According to triangular possibility distribution, Table 4.4 shows the assignment plans for UMOCOTS product selection problem-1 with different combinations of SP and AL.

4.6.1.1   *CONVERGENCE RATE OF GA FOR COTS SELECTION PROBLEM-1*

The convergence rate of GA for COTS selection problem-1 is obtained for case-1 of Table 4.3 at α = 0.1, 0.5, and 0.9 as shown in Figures 4.5–4.7. For the convergence rate, we used different number of iterations and populations with the values $\max W = \prod_{i=1}^{2} \prod_{j=1}^{3} \mu_{Z_{ij}}$.

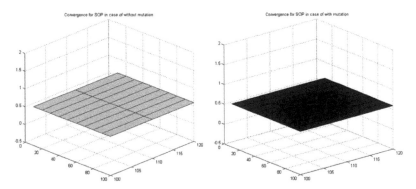

FIGURE 4.5   Convergence rate of GA for case-1 α = 0.1.

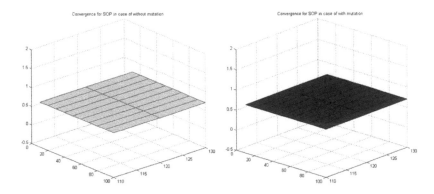

FIGURE 4.6   Convergence rate of GA for case-1at α = 0.5.

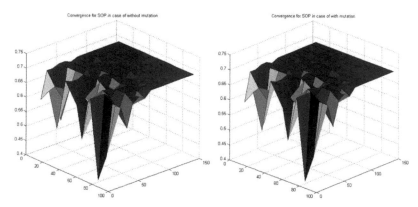

FIGURE 4.7 Convergence rate of GA for case-1at $\alpha = 0.9$.

In both cases, the GA-based algorithm converges after 100 populations and 10 iterations, 100 populations and 10 iterations as well as 120 populations and 70 iterations for case-1 of Table 4.3 at $\alpha = 0.1$ $\alpha = 0.5$ and $\alpha = 0.9$, respectively, which are shown in Figures 4.5–4.7. For other cases of Table 4.3, the convergence rate of GA approximately remains the same. Figure 4.7 also provided other solution options to DM as his/her need.

Figure 4.8 indicates that the ESs of quality and cost objectives as (0.8018, 0.8420, 0.8870) and (18.7, 34, 45.7) at $\alpha = 0.1$, (0.8259, 0.8461, 0.8753) and (25, 35, 40.5) at $\alpha = 0.5$, (0.8305, 0.8339, 0.8384) and (31.2, 33, 34.5) at $\alpha = 0.9$ for (−5, −5) SP and (0.8, 0.9) AL, respectively. Similarly, Figure 4.9 indicates the ESs of both the objectives (0.8018, 0.8420, 0.8870) and (18.7, 34, 45.7) at $\alpha = 0.1$, (0.8259, 0.8461, 0.8753) and (25, 35, 40.5) at $\alpha = 0.5$, (0.8410, 0.8446, 0.8490) and (32.4, 34, 35.7) at $\alpha = 0.9$ for (−1, −2) SP and (0.4, 0.65) AL, respectively.

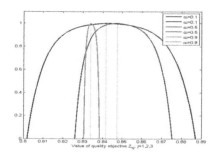
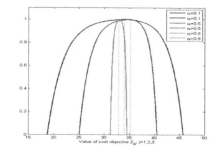

FIGURE 4.8 Possibilities distribution for quality and cost objectives for case-1.

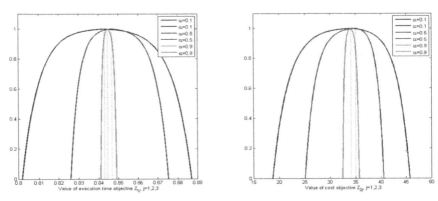

FIGURE 4.9 Possibilities distribution for quality and cost objectives for case-6.

Figure 4.10 indicates the degree of satisfaction levels (0.8752, 0.8656, 0.8021), (0.8752, 0.8656, 0.8021), (0.8752, 0.8656, 0.8021), (0.7419, 0.7286, 0.6478), (0.6440, 0.6294, 0.5450), and (0.5289, 0.5140, 0.4361) of the quality objective for three different scenario at (−5, −5), (−5, −1), (−5, −2), (−3, −5), (−2, −3), and (−1, −2) SPs and thier corrosponding ALs (0.8, 0.9), (0.75, 0.7), (0.8, 0.7), (0.6, 0.8), (0.5, 0.8), and (0.4, 0.65) for $\alpha = 0.1$. Similarly, Figure 4.11 shows the degree of satisfaction levels (1, 0.9709, 0.9987), (1, 0.7698, 0.9769), (1, 0.8517, 0.9888), (1, 0.9709, 0.9987), (1, 0.91, 0.9943), and (1, 0.8517, 0.9888) of the cost objective for three different scenario at (−5, −5), (−5, −1), (−5, −2), (−3, −5), (−2, −3), and (−1, −2) SPs and thier corrosponding ALs (0.8, 0.9), (0.75, 0.7), (0.8, 0.7), (0.6, 0.8), (0.5, 0.8), and (0.4, 0.65) for $\alpha = 0.1$. In general, the achievement levels may not be large enough to fulfill the DM due to the multiobjective characteristic of the problem.

4.6.2 PROBLEM-2: UMOCOTS PRODUCT SELECTION PROBLEM WITH TRAPEZOIDAL FUZZY NUMBERS

To demonstrate the appropriateness of the GA-based approach on COTS selection Model-2, the UMOCOTS selection problem has been referred from.[5,14] The different parameters of the COTS selection problem with trapezoidal fuzzy numbers are given in Table 4.5.

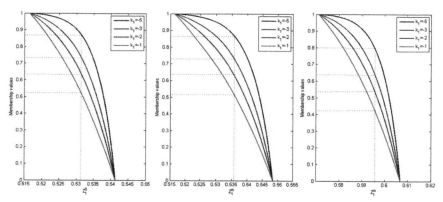

FIGURE 4.10 The degree of satisfaction of $Z_{1j}; j = 1,2,3$ at $\alpha = 0.1$.

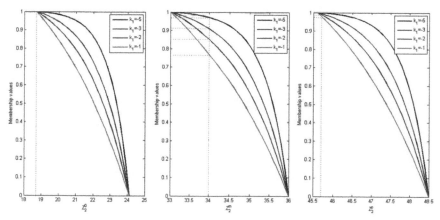

FIGURE 4.11 The degree of satisfaction of $Z_{2j}; j = 1,2,3$ at $\alpha = 0.1$.

From Table 4.5, the UMOCOTS selection problem model-2 is written as follows

$$\max Z_1 = (6,8,10,12)x_{11} + (7,9,12,15)x_{12} + (8,11,14,16)x_{13} + (7,8,11,13)x_{21} + (7,9,11,14)x_{22}$$
$$+ (6,9,11,12)x_{31} + (6,8,11,14)x_{32} + (5,7,9,11)x_{33} + (6,8,12,14)x_{41} + (8,10,12,14)x_{42} + (4,7,8,10)x_{43}$$

$$\min Z_2 = (14000,16000,19000,21000)x_{11} + (13000,15000,17000,19000)x_{12} + (8000,10000,12000,14000)x_{13} +$$
$$(11000,13000,14000,16000)x_{21} + (11000,12000,13000,14000)x_{22} + (3000,5000,7000,9000)x_{31} +$$
$$(4000,6000,7000,9000)x_{32} + (6000,8000,9000,11000)x_{33} + (9000,11000,13000,14000)x_{41} +$$
$$(11000,13000,15000,16000)x_{42} + (13000,15000,17000,19000)x_{43}$$

TABLE 4.5 The Different Parameters of COTS.

Cost

$c_{11} = (6,8,10,12)$ $\qquad$ $c_{21} = (7,8,11,13)$ $\qquad$ $c_{31} = (6,9,11,12)$ $\qquad$ $c_{41} = (6,8,12,14)$

$c_{12} = (7,9,12,15)$ $\qquad$ $c_{22} = (7,9,11,14)$ $\qquad$ $c_{32} = (6,8,11,14)$ $\qquad$ $c_{42} = (8,10,12,14)$

$c_{13} = (8,11,14,16)$ $\qquad\qquad\qquad\qquad$ $c_{33} = (5,7,9,11)$ $\qquad$ $c_{43} = (4,7,8,10)$

No. of lines of code

$l_{11} = (14000,16000,19000,21000)$ $\quad$ $l_{21} = (11000,13000,14000,16000)$ $\quad$ $l_{31} = (3000,5000,7000,9000)$ $\quad$ $l_{41} = (9000,11000,13000,14000)$

$l_{12} = (13000,15000,17000,19000)$ $\quad$ $l_{22} = (11000,12000,13000,14000)$ $\quad$ $l_{32} = (4000,6000,7000,9000)$ $\quad$ $l_{42} = (11000,13000,15000,16000)$

$l_{13} = (8000,10000,12000,14000)$ $\qquad\qquad\qquad\qquad\qquad\qquad$ $l_{33} = (6000,8000,9000,11000)$ $\quad$ $l_{43} = (13000,15000,17000,19000)$

Execution time

$t_{11} = (1.5,2.5,3.5,5)$ $\qquad$ $t_{21} = (1,1.5,2,3)$ $\qquad$ $t_{31} = (3,4.6,5,6)$ $\qquad$ $t_{41} = (6,8,9,10)$

$t_{12} = (1,1.6,2,2.5)$ $\qquad$ $t_{22} = (1.5,2.2,3,4)$ $\qquad$ $t_{32} = (2,3.5,4,5)$ $\qquad$ $t_{42} = (4,6,7,8)$

$t_{13} = (0.5,1,2,3)$ $\qquad\qquad\qquad\qquad$ $t_{33} = (4,4.6,5,6)$ $\qquad$ $t_{43} = (4,5,7,8)$

Delivery time

$d_{11} = (5,7,10,11)$ $\qquad$ $d_{21} = (5,7,10,12)$ $\qquad$ $d_{31} = (3,5,6,8)$ $\qquad$ $d_{41} = (3,5,7,10)$

$d_{12} = (3,4,6,8)$ $\qquad$ $d_{22} = (3,6,8,11)$ $\qquad$ $d_{32} = (3,7,8,10)$ $\qquad$ $d_{42} = (4,8,10,12)$

$d_{13} = (4,5,7,9)$ $\qquad\qquad\qquad\qquad$ $d_{33} = (2,4,6,8)$ $\qquad$ $d_{43} = (3,5,7,11)$

TABLE 4.5 *(Continued)*

Probability of use of program

| $v_1 = 0.411$ | $v_2 = 0.261$ | $v_3 = 0.328$ |

Probability of failure on demand

$\mu_{11} = 0.0005$	$\mu_{21} = 0.0004$	$\mu_{31} = 0.0006$	$\mu_{41} = 0.0006$
$\mu_{12} = 0.0003$	$\mu_{22} = 0.0002$	$\mu_{32} = 0.0004$	$\mu_{42} = 0.0003$
$\mu_{13} = 0.0002$		$\mu_{33} = 0.0002$	$\mu_{43} = 0.0008$

Average no. of invocations

| $S_1 = 80$ | $S_2 = 55$ | $S_3 = 120$ | $S_1 = 110$ |

Maximum desired level

$T = (3,9,10,12)$

Contingent decision constraints

$x_{42} \leq x_{11}$ or $x_{42} \leq x_{13}$

$$\max Z_1 = (6,8,10,12)x_{11} + (7,9,12,15)x_{12} + (8,11,14,16)x_{13} + (7,8,11,13)x_{21} + (7,9,11,14)x_{22}$$
$$+ (6,9,11,12)x_{31} + (6,8,11,14)x_{32} + (5,7,9,11)x_{33} + (6,8,12,14)x_{41} + (8,10,12,14)x_{42} + (4,7,8,10)x_{43}$$

$$\min Z_2 = (14000,16000,19000,21000)x_{11} + (13000,15000,17000,19000)x_{12} + (8000,10000,12000,14000)x_{13} +$$
$$(11000,13000,14000,16000)x_{21} + (11000,12000,13000,14000)x_{22} + (3000,5000,7000,9000)x_{31} +$$
$$(4000,6000,7000,9000)x_{32} + (6000,8000,9000,11000)x_{33} + (9000,11000,13000,14000)x_{41} +$$
$$(11000,13000,15000,16000)x_{42} + (13000,15000,17000,19000)x_{43}$$

$$\max Z_3 = (1.5,2.5,3.5,5)x_{11} + (1,1.6,2,2.5)x_{12} + (0.5,1,2,3)x_{13} + (1,1.5,2,3)x_{21} + (1.5,2.2,3,4)x_{22}$$
$$+ (3,4.6,5,6)x_{31} + (2,3.5,4,5)x_{32} + (4,4.6,5,6)x_{33} + (6,8,9,10)x_{41} + (4,6,7,8)x_{42} + (4,5,7,8)x_{43}$$

subject to the constraints:

$$(5,7,10,12)x_{11} + (3,4,6,8)x_{12} + (4,5,7,9)x_{13} \le (3,9,10,12),$$

$$(5,7,10,12)x_{21} + (3,6,8,11)x_{22} \le (3,9,10,12),$$

$$(3,5,6,8)x_{31} + (3,7,8,10)x_{32} + (2,4,6,8)x_{33} \le (3,9,10,12),$$

$$(3,5,7,10)x_{41} + (4,8,10,12)x_{42} + (3,5,7,11)x_{43} \le (3,9,10,12),$$

$$\exp(-80(0.0005x_{11} + 0.0003x_{12} + 0.0002x_{13}) \times \exp(-55(0.0004x_{21} + 0.0002x_{22}) \times$$
$$\exp(-120(0.0006x_{31} + 0.0004x_{32} + 0.0002x_{33}) \times \exp(-110(0.0006x_{41} + 0.0003x_{42} + 0.0008x_{43}) \ge R ,$$

$$x_{11} + x_{12} + x_{13} = 1,$$

$$x_{21} + x_{22} = 1,$$

$$x_{31} + x_{32} + x_{33} = 1,$$

$$x_{41} + x_{42} + x_{43} = 1,$$

$$x_{21} - x_{11} \le 5y_1,$$

$$x_{42} - x_{13} \le 5y_2,$$

$$y_1 + y_2 = 1,$$

$$y_1, y_2 \in \{0,1\},$$

$$x_{ij} \in \{0,1\}, \forall i, j.$$

To evaluate the UMOCOTS product selection problem, the model is coded.[13] For COTS selection problem-2, the PIS and NIS for Z_{ij}; $i = 1,2,3$ and $j = 1,2,3,4$ of model-2 at $\alpha = 0.1, 0.5$ and 0.9 are shown in Table 4.6

TABLE 4.6 PIS and NIS for Z_{ij}, $i = 1,2,3$ and $j = 1,2,3,4$ at $\alpha = 0.1$, 0.5 and 0.9.

α-level	Solutions	Objectives											
		Z_{11}	Z_{12}	Z_{13}	Z_{14}	Z_{21}	Z_{22}	Z_{23}	Z_{24}	Z_{31}	Z_{32}	Z_{33}	Z_{34}
$\alpha = 0.1$	PIS	23.9	32	40	49	36,700	43,000	50,000	55,400	4.4495	6.3623	8.1460	9.9086
	NIS	26.9	35	46	55.9	43,700	50,000	56,000	62,300	6.2633	7.9942	9.3910	11.1536
$\alpha = 0.5$	PIS	26	30	38	42	34,500	38,000	45,000	48,000	4.49	5.4485	7.4740	8.45325
	NIS	34	39	48	53	48,000	52,000	59,000	63,000	7.5499	8.6593	10.4995	11.84825
$\alpha = 0.9$	PIS	29.2	30	38	38.8	37,300	38,000	45,000	45,600	5.2568	5.4485	7.4740	7.66985
	NIS	.38	39	48	49	51,200	52,000	59,000	59,800	8.43742	8.6593	10.4995	10.76925

To convert the UMOCOTS product selection problem-2 into its crisp equivalent MOOM-2.1, we used the minimum desired level of excepted reliability R=0.82.[5,14]

Each grouping of the SPs and its corresponding different ALs are shown in Table 4.7.

TABLE 4.7   Each Combination of SPs and AL for COTS Selection Problem-2.

Case	Shape parameters	Aspiration level
1	$(-5, -5, -5)$	0.8, 0.75, 0.85
2	$(-5, -5, -5)$	0.9, 0.7, 0.95
3	$(-1, -2, -5)$	0.65, 0.5, 0.8
4	$(-3, -5, -2)$	0.85, 0.6, 0.7
5	$(-2, -3, -1)$	0.7, 0.55, 0.8
6	$(-5, -3, -2)$	0.85, 0.6, 0.5
7	$(-1, -1, -1)$	0.5, 0.35, 0.7

According to the trapezoidal possibility distribution, Table 4.8 shows the assignment plans for UMOCOTS product selection problem-2 with different combinations of SP and AL. In Tables 4.4 and 4.8, the values of $\alpha = 0.1, 0.5$ and 0.9 are taken to replicate the different circumstances of DM's assurance on fuzzy decision. It is clear from Tables 4.4 and 4.8 that the variation in α directly affect all objectives.[4,10]

4.6.2.1   CONVERGENCE RATE OF GA FOR PROBLEM-2

The convergence rate of GA for COTS selection problem-2 is obtained for case-1 of Table 4.7 at $\alpha = 0.1, 0.5$, and 0.9 as shown in Figures 4.12–4.14. For the convergence rate, we used different number of iterations and populations with the values of $\max W = \prod_{i=1}^{2}\prod_{j=1}^{3}\mu_{Z_{ij}}$.

TABLE 4.8 Summary Results of Problem-2 at $\alpha = 0.1, 0.5, 0.9$.

α	Case	Degree of satisfaction	Membership values $(\mu_{Z_{1j}}, \mu_{Z_{2j}}, \mu_{Z_{3j}})$	Objective values Z_1, Z_2, Z_3	Optimum allocations
$\alpha =$ 0.1	1	0.7655	(0.9709, 0.9709, 0.9709, 0.9531) (0.7655, 0.7655, 0.8166, 0.7705) (1, 1, 1, 1)	(24.9, 33, 42, 51.9) (41700, 48000, 54000, 60300) (4.4495, 6.3623, 8.1460, 9.9087)	$SC_{12}, SC_{22}, SC_{32}, SC_{43}$
	2	0.7655	(0.9709, 0.9709, 0.9709, 0.9531) (0.7655, 0.7655, 0.8166, 0.7705) (1, 1, 1, 1)	(24.9, 33, 42, 51.9) (41700, 48000, 54000, 60300) (4.4495, 6.3623, 8.1460, 9.9087)	$SC_{12}, SC_{22}, SC_{32}, SC_{43}$
	3	0.5034	(0.7698, 0.7698, 0.7698, 0.6960) (0.5034, 0.5034, 0.5627, 0.5088) (1, 1, 1, 1)	(24.9, 33, 42, 51.9) (41700, 48000, 54000, 60300) (4.4495, 6.3623, 8.1460, 9.9087)	$SC_{12}, SC_{22}, SC_{32}, SC_{43}$
	4	0.7655	(0.9100, 0.9100, 0.9100, 0.8675) (0.7655, 0.7655, 0.8166, 0.7705) (1, 1, 1, 1)	(24.9, 33, 42, 51.9) (41700, 48000, 54000, 60300) (4.4495, 6.3623, 8.1460, 9.9087)	$SC_{12}, SC_{22}, SC_{32}, SC_{43}$
	5	0.6058	(0.8517, 0.8517, 0.8517, 0.7938) (0.6058, 0.6058, 0.6652, 0.6113) (1, 1, 1, 1)	(24.9, 33, 42, 51.9) (41700, 48000, 54000, 60300) (4.4495, 6.3623, 8.1460, 9.9087)	$SC_{12}, SC_{22}, SC_{32}, SC_{43}$

TABLE 4.8 (Continued)

α	Case	Degree of satisfaction	Membership values $(\mu_{z1j}, \mu_{z2j}, \mu_{z3j})$	Objective values Z_1, Z_2, Z_3	Optimum allocations
	6	0.6058	(0.9709, 0.9709, 0.9709, 0.9513) (0.6058, 0.6058, 0.6652, 0.6113) (1, 1, 1, 1)	(24.9, 33, 42, 51.9) (41700, 48000, 54000, 60300) (−4.4495, 6.3623, 8.1460, 9.9087)	$sc_{12}, sc_{22}, sc_{32}, sc_{43}$
	7	0.3932	(0.7698, 0.7698, 0.7698, 0.6960) (0.3932, 0.3932, 0.4484, 0.3981) (1, 1, 1, 1)	(24.9, 33, 42, 51.9) (41700, 48000, 54000, 60300) (−4.4495, 6.3623, 8.1460, 9.9087)	$sc_{12}, sc_{22}, sc_{32}, sc_{43}$
$\alpha = 0.5$	1	0.8705	(0.9463, 0.9442, 0.8705, 0.8766) (0.9548, 0.9490, 0.9663, 0.9567) (1, 1, 0.9979, 1)	(29.5, 34, 44, 48.5) (40000, 44000, 50000, 54000) (−4.49, 5.4485, 7.4740, 8.6380)	$sc_{13}, sc_{21}, sc_{32}, sc_{43}$
	2	0.8887	(0.9744, 0.9709, 0.9567, 0.9650) (0.8977, 0.8887, 0.9241, 0.9092) (0.9766, 0.9882, 0.9888, 0.9856)	(28.5, 33, 42, 46) (42000, 46000, 52000, 56000) (5.4030, 6.0964, 8.0630, 9.2270)	$sc_{13}, sc_{21}, sc_{33}, sc_{43}$
	3	0.6657	(0.7865, 0.7698, 0.7138, 0.7448) (0.6811, 0.6657, 0.7311, 0.7017) (0.9766, 0.9882, 0.9888, 0.9856)	(28.5, 33, 42, 46) (42000, 46000, 52000, 56000) (5.4030, 6.0964, 8.0630, 9.2270)	$sc_{13}, sc_{21}, sc_{33}, sc_{43}$
	4	0.8723	(0.9186, 0.9100, 0.8784, 0.8964) (0.8977, 0.8887, 0.9241, 0.9092) (0.8723, 0.9222, 0.9255, 0.9096)	(28.5, 33, 42, 46) (42000, 46000, 52000, 56000) (5.4030, 6.0964, 8.0630, 9.2270)	$sc_{13}, sc_{21}, sc_{33}, sc_{43}$

TABLE 4.8 (Continued)

α	Case	Degree of satisfaction	Membership values $(\mu_{Z1j}, \mu_{Z2j}, \mu_{Z3j})$	Objective values Z_1, Z_2, Z_3	Optimum allocations
	5	0.5689	(0.8641, 0.8517, 0.8082, 0.7311) (0.5689, 0.6058, 0.6919, 0.7021) (0.8237, 0.8084, 0.8553, 0.8726)	(28.5, 33, 42, 47.5) (44500, 48000, 54000, 57500) (5.2997, 6.3623, 8.1460, 9.1253)	$SC_{12}, SC_{22}, SC_{32}, SC_{43}$
	6	0.8629	(0.9463, 0.9442, 0.8705, 0.8766) (0.8745, 0.8629, 0.8994, 0.8784) (1, 1, 1, 0.9820)	(29.5, 34, 44, 48.5) (40000, 44000, 50000, 54000) (4.49, 5.4485, 7.4740, 8.6380)	$SC_{13}, SC_{21}, SC_{32}, SC_{43}$
	7	0.5215	(0.6806, 0.6743, 0.5215, 0.5311) (0.7073, 0.6886, 0.7502, 0.7138) (1, 1, 1, 0.9675)	(29.5, 34, 44, 48.5) (40000, 44000, 50000, 54000) (4.49, 5.4485, 7.4740, 8.6380)	$SC_{13}, SC_{21}, SC_{32}, SC_{43}$
$\alpha = 0.9$	1	0.8705	(0.9446, 0.9442, 0.8705, 0.8719) (0.9501, 0.9490, 0.9663, 0.9645) (1, 1, 1, 0.9996)	(33.1, 34, 44, 44.9) (43200, 44000, 50000, 50800) (5.2568, 5.4485, 7.4740, 7.7068)	$SC_{13}, SC_{21}, SC_{32}, SC_{43}$
	2	0.8887	(0.9715, 0.9709, 0.9567, 0.9586) (0.8905, 0.8887, 0.9241, 0.9212) (0.9864, 0.9882, 0.9888, 0.9882)	(32.1, 33, 42, 42.8) (45200, 46000, 52000, 52800) (5.9577, 6.0964, 8.0630, 8.2958)	$SC_{13}, SC_{21}, SC_{33}, SC_{43}$
	3	0.6657	(0.7728, 0.7698, 0.7138, 0.7206) (0.6687, 0.6657, 0.7311, 0.7250) (0.9864, 0.9882, 0.9888, 0.9882)	(32.1, 33, 42, 42.8) (45200, 46000, 52000, 52800) (5.9577, 6.0964, 8.0630, 8.2958)	$SC_{13}, SC_{21}, SC_{33}, SC_{43}$

TABLE 4.8 *(Continued)*

α Case	Degree of satisfaction	Membership values $(\mu_{Z1j}, \mu_{Z2j}, \mu_{Z3j})$	Objective values Z_1, Z_2, Z_3	Optimum allocations
4	0.8784	(0.9116, 0.9100, 0.8784, 0.8825)	(32.1, 33, 42, 42.8)	$SC_{13}, SC_{21}, SC_{33}, SC_{43}$
		(0.8905, 0.8887, 0.9241, 0.9212)	(45200, 46000, 52000, 52800)	
		(0.9133, 0.9222, 0.9255, 0.9211)	(5.9577, 6.0964, 8.0630, 8.2958)	
5	0.7615	(0.8540, 0.8517, 0.8082, 0.8136)	(32.1, 33, 42, 42.8)	$SC_{13}, SC_{21}, SC_{33}, SC_{43}$
		(0.7641, 0.7615, 0.8176, 0.8126)	(45200, 46000, 52000, 52800)	
		(0.8656, 0.8699, 0.8749, 0.8698)	(5.9577, 6.0964, 8.0630, 8.2958)	
6	0.8652	(0.9446, 0.9442, 0.8705, 0.8719)	(33.1, 34, 44, 44.9)	$SC_{13}, SC_{21}, SC_{32}, SC_{43}$
		(0.8652, 0.8929, 0.8994, 0.8952)	(43200, 44000, 50000, 50800)	
		(1, 1, 1, 0.9962)	(5.2568, 5.4485, 7.4740, 7.7068)	
7	0.5215	(0.6755, 0.6743, 0.5215, 0.5236)	(33.1, 34, 44, 44.9)	$SC_{13}, SC_{21}, SC_{32}, SC_{43}$
		(0.6923, 0.6886, 0.7502, 0.7426)	(43200, 44000, 50000, 50800)	
		(1, 1, 1, 0.9930)	(5.2568, 5.4485, 7.4740, 7.7068)	

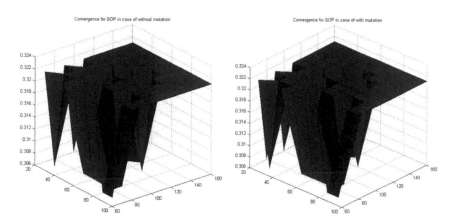

FIGURE 4.12   Convergence rate of GA for case-1at $\alpha = 0.1$.

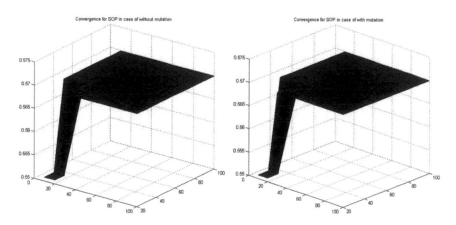

FIGURE 4.13   Convergence rate of GA for case-1at $\alpha = 0.5$.

In the without and with mutation operator cases, the GA-based algorithm converges after 130 populations and 80 iterations, 30 populations and 40 iterations as well as 40 populations and 10 iterations for case-1 of Table 4.7 at $\alpha = 0.1$ $\alpha = 0.5$ and $\alpha = 0.9$, respectively, which are shown in Figures 4.12–4.14. For other cases of Table 4.7, the convergence rate of GA approximately remains the same. Figures 4.12 and 4.13 also provided other solution options to DM as his/her need.

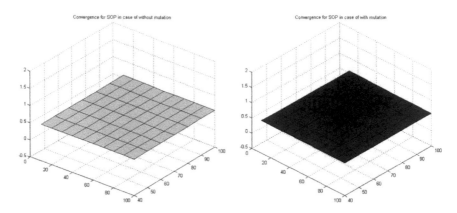

FIGURE 4.14   Convergence rate of GA for case-1at $\alpha = 0.9$.

Figure 4.15 shows the ESs of cost, size, and execution time objectives as (24.9, 33, 42, 51.9), (41,700, 48,000, 54,000, 60,300), and (4.4495, 6.3623, 8.1460, 9.9087) at $\alpha = 0.1$, (29.5, 34, 44, 48.5), (40,000, 44,000, 50,000, 54,000), and (4.49, 5.4485, 7.4740, 8.6380) at $\alpha = 0.5$, (33.1, 34, 44, 44.9), (43,200, 44,000, 50,000, 50,800), and (5.2568, 5.4485, 7.4740, 7.7068) at $\alpha = 0.9$ for (−5, −5, −5) SP and (0.8, 0.75, 0.85) AL. Similar, Figure 4.16 provides the ESs of all objectives as (24.9, 33, 42, 51.9), (41,700, 48,000, 54,000, 60,300), and (4.4495, 6.3623, 8.1460, 9.9087) at$\alpha = 0.1$, (29, 34, 42, 46.5), (41,500, 45,000, 51,000, 54,500), and (5.8062, 6.5668, 8.7350, 9.8990) at $\alpha = 0.5$, (32.1, 33, 42, 42.8), (45,200, 46,000, 52,000, 52,800), and (5.9577, 6.0964, 8.0630, 8.2958) at $\alpha = 0.9$ for (−3, −5, −2) SP and (0.85, 0.6, 0.7) AL.

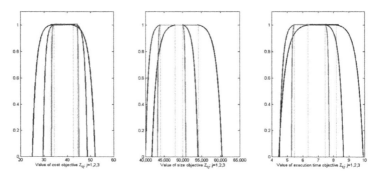

FIGURE 4.15   Possibilities distribution for cost, size, and execution time objectives for case-1.

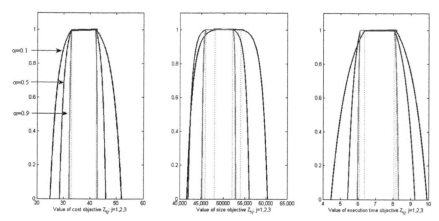

FIGURE 4.16 Possibilities distribution for cost, size, and execution time objectives for case-2.

From Figures 4.8, 4.9, 4.15 and 4.16, the effect of uncertainty on fuzzy opinion of DM is decreased and obtained solution has more effect of optimism rather than pessimism. The achieved solutions and optimum allocations in Tables 4.4 and 4.11 show the best advantage to use the EMF with a variety of SPs in the UMOCOTS product selection problems. If the DM is not fulfilled with accepting the assignment plan, further plans can be made by altering the values of α and the SP.

Figure 4.17 shows the degree of satisfaction levels (0.9709, 0.9709, 0.9709, 0.9531), (0.9709, 0.9709, 0.9709, 0.9531), (0.7698, 0.7698, 0.7698, 0.6960), (0.9100, 0.9100, 0.9100, 0.8675), (0.8517, 0.8517, 0.8517, 0.7938), (0.9709, 0.9709, 0.9709, 0.9513), and (0.7698, 0.7698, 0.7698, 0.6960) of cost objective for the four different scenarios at $(-5, -5, -5)$, $(-5, -5, -5)$, $(-1, -2, -5)$, $(-3, -5, -2)$, $(-2, -3, -1)$, $(-5, -3, -2)$, and $(-1, -1, -1)$ SPs and their corresponding (0.8, 0.75, 0.85), (0.9, 0.7, 0.95), (0.65, 0.5, 0.8), (0.85, 0.6, 0.7), (0.7, 0.55, 0.8), (0.85, 0.6, 0.5), and (0.5, 0.35, 0.7) ALs for $\alpha = 0.1$. Similar Figures 4.18 and 4.19 show the degree of satisfaction levels (0.7655, 0.7655, 0.8166, 0.7705), (0.7655, 0.7655, 0.8166, 0.7705), (0.5034, 0.5034, 0.5627, 0.5088), (0.7655, 0.7655, 0.8166, 0.7705), (0.6058, 0.6058, 0.6652, 0.6113), (0.6058, 0.6058, 0.6652, 0.6113), and (0.3932, 0.3932, 0.4484, 0.3981) of size objective and (1, 1, 1), (1, 1, 1), (1, 1, 1), (1, 1, 1), (1, 1, 1), (1, 1, 1), and (1, 1, 1) of execution time objective at $(-5, -5, -5)$, $(-5, -5, -5)$, $(-1, -2, -5)$, $(-3, -5, -2)$, $(-2, -3, -1)$, $(-5, -3, -2)$, and $(-1, -1, -1)$ SPs and their corresponding (0.8, 0.75, 0.85), (0.9, 0.7, 0.95), (0.65, 0.5, 0.8), (0.85, 0.6, 0.7), (0.7, 0.55, 0.8), (0.85, 0.6, 0.5), and (0.5, 0.35, 0.7) ALs for $\alpha = 0.1$.

As shown in Tables 4.4 and 4.8 and Figures 4.10, 4.11, 4.17–4.19, the GA-based approach provided flexibility and large data collection in the sense of altering SPs. It provides an analysis of different situations to DM for assignment strategy in UMOCOTS selection problems. So, different solutions are chosen by DM in different situations according to DM's necessity Depending upon the predilections of the DM for all objectives, the preferred compromise solution also can be adapted by alternating the α, SP's values as well as the AL's values.[4,10]

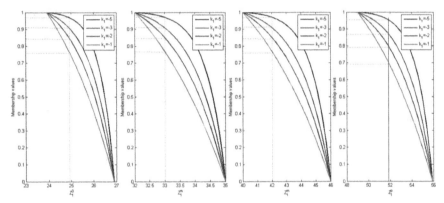

FIGURE 4.17   The degree of satisfaction of Z_{1j}; $j = 1,2,3,4$ at $\alpha = 0.1$.

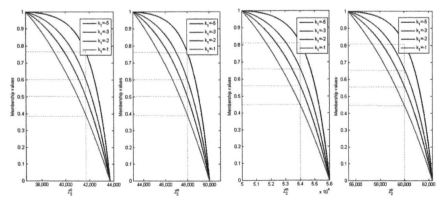

FIGURE 4.18   The degree of satisfaction of Z_{2j}; $j = 1,2,3,4$ at $\alpha = 0.1$.

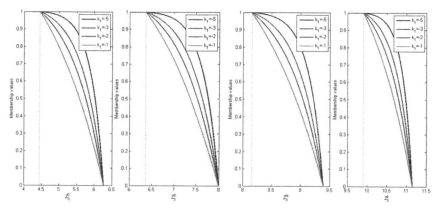

FIGURE 4.19 The degree of satisfaction of Z_{3j}; $j = 1,2,3,4$ at $\alpha = 0.1$.

4.7 COMPARISON

To more assessment of the developed approach, we compare the obtained results with other approaches. Tables 4.9 and 4.10 show a comparison of the solutions which are obtained by GA-based approach using possibility distribution with other approaches for UMOCOTS product selection problem-1 and problem-2, respectively.

TABLE 4.9 Comparison of GA-Based Approach with Different Approaches for Problem-1.

Degree of satisfaction levels and Objective values		SP and AL	Degree of satisfaction levels and Objective values
Fuzzy interactive approach[9]	Pankaj Gupta et al. approach[6]		GA-based hybrid approach with possibility distribution at $\alpha = 0.1$
(0.871533, 34)	**0.46245**	(−5, −5)	**0.8021**
	(0.8565, 0.84612, 0.90442),	(0.8, 0.9)	(0.8018, 0.8420, 0.8870), (18.7, 34, 45.7)
	(15, 35, 46)		
		(−5, −1)	**0.7698**
		(0.75, 0.7)	(0.8018, 0.8420, 0.8870), (18.7, 34, 45.7)

TABLE 4.9   *(Continued)*

(−5, −2)	**0.8021**
(0.8, 0.7)	(0.8018, 0.8420, 0.8870), (18.7, 34, 45.7)
(−3, −5)	**0.6478**
(0.6, 0.8)	(0.8018, 0.8420, 0.8870), (18.7, 34, 45.7)
(−2, −3)	**0.5450**
(0.5, 0.8)	(0.8018, 0.8420, 0.8870), (18.7, 34, 45.7)
(−1, −2)	**0.4361**
(0.4, 0.65)	(0.8018, 0.8420, 0.8870), (18.7, 34, 45.7)

TABLE 4.10   Comparison of GA-Based Approach with Different Approaches for Problem-2.

Weights	Degree of satisfaction level and objective values		
	Two phase approach[15]	Weighted approach[19]	Pankaj, Mukesh approach[13]
(0.1, 0.2, 0.7)	**0.52631**	**0.05405**	**0.3750**
	(37.75, 48,000, 7.8526)	(42.75, 44,750, 6.646)	(39, 47,000, 6.564)
(0.1, 0.7, 0.2)	**0.52631**	**0.0313**	**0.0313**
	(37.75, 48,000, 7.8526)	(41.75, 42,750, 7.302)	(41.75, 42,750, 7.302)
(0.7, 0.1, 0.2)	**0.52631**	**0.000**	**0.1053**
	(37.75, 48,000, 7.8526)	(34, 55,500, 8.4235)	(35.75, 54,000, 7.229)

SP and AL	Degree of satisfaction levels and objective values
	GA-based hybrid approach with possibility distribution at $\alpha = 0.1$
(−5, −5, −5)	**0.7655**
(0.8, 0.75, 0.85)	(24.9, 33, 42, 51.9), (41,700, 48,000, 54,000, 60,300), (4.4495, 6.3623, 8.1460, 9.9087)
(−5, −5, −5)	**0.7655**
(0.9, 0.7, 0.95)	(24.9, 33, 42, 51.9), (41,700, 48,000, 54,000, 60,300), (4.4495, 6.3623, 8.1460, 9.9087)
(−1, −2, −5)	**0.5034**
(0.65, 0.5, 0.8)	(24.9, 33, 42, 51.9), (41,700, 48,000, 54,000, 60,300), (4.4495, 6.3623, 8.1460, 9.9087)

TABLE 4.10   *(Continued)*

(−3, −5, −2)	**0.7655**
(0.85, 0.6, 0.7)	(24.9, 33, 42, 51.9), (41,700, 48,000, 54,000, 60,300), (4.4495, 6.3623, 8.1460, 9.9087)
(−2, −3, −1)	**0.6058**
(0.7, 0.55, 0.8)	(24.9, 33, 42, 51.9), (41,700, 48,000, 54,000, 60,300), (4.4495, 6.3623, 8.1460, 9.9087)
(−5, −3, −2)	**0.6058**
(0.85, 0.6, 0.5)	(24.9, 33, 42, 51.9), (41,700, 48,000, 54,000, 60,300), (4.4495, 6.3623, 8.1460, 9.9087)
(−1, −1, −1)	**0.3932**
(0.5, 0.35, 0.7)	(24.9, 33, 42, 51.9), (41,700, 48,000, 54,000, 60,300), (4.4495, 6.3623, 8.1460, 9.9087)

Tables 4.9 and 4.10 indicate that GA-based hybrid approach with possibility distribution can effectively handle vagueness and imprecision in the objective values with triangular and trapezoidal fuzzy numbers of COTS selection problems. GA-based approach with possibility distribution keeps the uncertainty in problem until the solution is reached hence, this approach provides the different scenarios as the optimistic, the most-likely, and the pessimistic scenarios for all uncertain objective functions and also provide a lot of information about the solutions.[4,11,23] Furthermore, GA-based approach with possibility theory can solve rather general inexact problems throughout an interactive process with DM and also achieved the complete possibility distribution to satisfy objective value.

4.8   CONCLUSION

GA-based approach with possibility distribution rendered the solution of the UMOCOTS product selection problems with respect to many choices of SPs and their corresponding ALs. The EMF is used to effectively design the SONOP with the product operator. The SPs in the EMFs are used to depicted classes of accuracy in objective functions. The consequential SONOP with some resource constraint is "NP-hard" and the various alternatives of ALs desired for the objectives specified by the DM, so it is worked out with GA. This approach describes how the different parameters affect the algorithm and their effect on convergence to concluding

ES. This approach also provided distinct fuzzy utilities of the DM and the performance of uncertainty in best-fit COTS selection.

KEYWORDS

- **multiobjective optimization**
- **COTS selection**
- **possibility distribution**
- **α–level set**
- **genetic algorithm**

REFERENCES

1. Berman, O.; Ashrafi, N. Optimization Models for Reliability of Modular Software Systems. *IEEE Trans. Softw. Eng.* **1993**,*19* (11), 1119–1123.
2. Buckley, J. Possibilistic Linear Programming with Triangular Fuzzy Numbers. *Fuzzy Sets Syst.* **1988**, *26* (1), 135–138.
3. Chi, D-H.; Lin, H-H.; Kuo, W. Software Reliability and Redundancy Optimization. *Reliability and Maintainability Symposium*, 1989. *Proceedings, Annual IEEE*, 1989.
4. Dhodiya, J. M.; Tailor, A. R. Genetic Algorithm Based Hybrid Approach to Solve Fuzzy Multi-Objective Assignment Problem Using Exponential Membership Function. *Springer Plus* Dec 1, **2016**, *5* (1), 2028.
5. Dhodiya, J. M.; Tailor, A. R. Genetic Algorithm Based Hybrid Approach to Solve Uncertain Multi-Objective COTS Selection Problem for Modular Software System. *J. Intell. Fuzzy Syst* **2018**, *34* (4), 2103–2120.
6. Gupta, P. et al. A Fuzzy Optimization Framework for COTS Products Selection of Modular Software Systems. *Int. J. Fuzzy Syst.* **2013**, *15*, 91–109.
7. Gupta, P. et al. A Hybrid Approach for Selecting Optimal COTS Products. In *International Conference on Computational Science and Its Applications*; Springer: Berlin Heidelberg, 2009.
8. Gupta, P.; Mehlawat, M. K.; Mittal, G. A Fuzzy Approach to Multicriteria Assignment Problem Using Exponential Membership Functions. *Int. J. Mach. Learn. Cybern.* **2013**, *4* (6), 647–657.
9. Gupta, P.; Mehlawat, M. K.; Verma, S. COTS Selection Using Fuzzy Interactive Approach. *Optim. Lett.* **2012**, *6* (2), 273–289.
10. Gupta, P.; Verma, S.; Mehlawat, M. K. A Membership Function Approach for Cost-Reliability Trade-Off of COTS Selection in Fuzzy Environment. *Int. J. Reliab. Qual. Saf. Eng.* **2011**, *18* (06), 573–595.

11. Gupta, P.; Mehlawat, M. K. A New Possibilistic Programming Approach for Solving Fuzzy Multiobjective Assignment Problem. *IEEE Trans. Fuzzy Syst.* **2014,** *22* (1), 16–34.

12. Jung, H-W.; Choi, B. Optimization Models for Quality and Cost of Modular Software Systems. *Eur. J. Oper. Res.* **1999,** *112* (3), 613–619.

13. Knowg, C. K. et al. Optimization of Software Components Selection for Component Based Software System Development. *Comput. Indust. Eng.* **2010,** *58* (4), 618–624.

14. Mehlawat, M. K.; Gupta, P. Multiobjective Credibilistic Model for COTS Products Selection of Modular Software Systems under Uncertainty. *Appl. Intell.* **2015,** *42* (2), 353–368.

15. Mehlawat, M. K. A Fuzzy Approach to Multiobjective COTS Products Selection of Modular Software Systems Using Exponential Membership Functions. *Int. J. Reliab. Qual. Saf. Eng.* **2014,** *21* (01), 1450005.

16. Mohamed, A.; Ruhe, G.; Eberlein, A. Cots Selection: Past, Present, and Future. In *Engineering of Computer-Based Systems, 2007. ECBS'07. 14th Annual IEEE International Conference and Workshops on the*; IEEE, 2007; pp 103–114.

17. Nakayama, H. Aspiration Level Approach to Interactive Multi-Objective Programming and Its Applications. In *Advances in Multicriteria Analysis*; Springer, **1995**; pp 147–174.

18. Nakayama, H.; UN, Y.; Yoon, M. Interactive Programming Methods for Multi-Objective Optimization. In *Sequential Approximate Multiobjective Optimization Using Computational Intelligence*, 2009; pp 17–43.

19. Neubauer, T.; Stummer, C. Interactive Decision Support for Multiobjective COTS Selection. In *System Sciences, 2007. HICSS 2007. 40th Annual Hawaii International Conference on*; IEEE, 2007.

20. Rajan, K. Adaptive Techniques in Genetic Algorithm and Its Applications, PhD thesis, Kottayam, 2013.

21. Rommelfanger, H. Interactive Decision Making in Fuzzy Linear Optimization Problems. *Eur. J. Oper. Res.* **1989,** *41* (2), 210–217.

22. Rommelfanger, H.; Anuscheck, R.; Wolf, J. Linear Programming with fuzzy Objectives. *Fuzzy Sets Syst.* **1989,** *29* (1), 31–48.

23. Sivanandam, S.; Deepa, S. *Introduction to Genetic Algorithms*; Springer Science & Business Media, 2007.

24. Shen, X.; Chen, Y.; Xing, L. Fuzzy Optimization Models for Quality and Cost of Software Systems Based on COTS. In *Proceedings of the Sixth International Symposium on Operations Research and Its Applications, Xinjiang ORSC & APORC*, Vol. 312318, Aug 2006.

25. Tailor, A. R.; Dhodiya, J. M. A Genetic Algorithm based Hybrid Approach to Solve Multi-objective Interval Assignment Problem by Estimation Theory. *Indian J. Sci. Technol.* Sept 29, **2016,** *9* (35).

26. Tailor AR, Dhodiya JM. Genetic Algorithm Based Hybrid Approach to Solve Multi-Objective Assignment Problem. *Int. J. Innov. Res. Sci. Eng. Technol.* **2016,** *5* (1), 524–535.

27. Tailor, A. R.; Dhodiya, J. M. Genetic Algorithm Based Hybrid Approach to Solve Optimistic, Most-Likely and Pessimistic Scenarios of Fuzzy Multi-Objective

Assignment Problem Using Exponential Membership Function. *Br. J. Math. Comput. Sci.* **2016,** *17* (2), 1–9.

28. Verma, S.; Mehlawat, M. K.; Mahajan, D. Software Component Evaluation and Selection Using TOPSIS and Fuzzy Interactive Approach Under Multiple Applications Development. *Ann. Oper. Res.* **2018,** 1–31.

29. Zadeh, L. A. Fuzzy Sets. *Info. Control* **1965,** *8* (3), 338–353.

30. Zahedi, F.; Ashrafi, N. Software Reliability Allocation Based on Structure, Utility, Price, and Cost. *IEEE Trans. Softw. Eng.* **1991,** *17* (4), 345–356.

31. Patel, K. K. R.; Tailor, A. R.; Patel, M.; Dhodiya, J. M. Fuzzy Theory Based Resource Allocation Problem with Possibilistic Approach: A Case Study of Sandwich Factory. *Int. J. Adv. Res. Ideas Innov. Technol.* **2018,** *4* (3).

32. Mehlawat, M. K.; Gupta, P. COTS Products Selection Using Fuzzy Chance-Constrained Multiobjective Programming. *Appl. Intell.* **2015,** *43* (4), 732–751.

33. Ketankumar, R. R.; Dhodiya, J. M. Possibilistic Distribution for Selection of Critical Path in Multi Objective Multi-Mode Problem with Trapezoidal Fuzzy Number. *Int. J. Recent Technol. Eng. (IJRTE)* **2019,** *8* (4).

CHAPTER 5

Fuzzy Logic Based Computational Technique for Analyzing Software Bug Repository

RAMA RANJAN PANDA* and NARESH KUMAR NAGWANI

*Department of Computer Science and Engineering,
National Institute of Technology, Raipur, Chhattisgarh, India*

Corresponding author. E-mail: rrpanda.phd2018.cs@nitrr.ac.in

ABSTRACT

Software development is a collaborative process in which programmers build software by integrating all the stages of the software development life cycle (SDLC). A software repository is a central file storage location where various software packages are stored, and these packages are retrieved and shared between all the software development team members at various locations. The software repositories are divided into various categories based on cooperation, coordination, and communication among the stakeholders as well as evolutionary changes to various software artifacts such as source code repositories, software bug repositories, historical repositories, run-time repositories and requirement documents, and other documentation. The software bug repository is an essential repository among the entire repositories since the completion of the software is entirely dependent on the bug fixing mechanism associated with this repository in software development. Today's software systems are larger and more complex as they go through various stages from the requirement

Computational Intelligence Applications for Software Engineering Problems. Parma Nand, PhD, Rakesh Nitin, PhD, Arun Prakash Agrawal, PhD & Vishal Jain, PhD (Eds.)
© 2023 Apple Academic Press, Inc. Co-published with CRC Press (Taylor & Francis)

analysis phase to the maintenance phase. A variety of tasks and activities are carried out in each stage of software development, and these are expensive and vulnerable to errors. During software development, a large number of software bugs are continuously generated, and that has become the main reason for the delay in software completion. Hence, there is a vast demand for computational intelligent techniques to accomplish various tasks of software development. In recent years, fuzzy logic techniques emerged and played an important role in various fields of data mining and text mining. Since most of the content related to software bug repositories is text in nature, it is possible to effectively use fuzzy logic techniques to analyze these software bugs.

5.1   INTRODUCTION

The software industry uses a software development life cycle (SDLC) process to model, build, and evaluate high-quality software. The main objective of SDLC is to develop quality software according to the customer's specifications at the lowest cost in the pre-defined time frame.[6,7] A typical SDLC consists of phases from requirement analysis to maintenance. SDLC provides a well-structured flow of phases that help the organization to easily build high-quality software that is well-tested and ready for production use. Every phase of the SDLC life cycle has its own set of processes and deliverables that flow into the next phase. Today's software systems are larger and more complex as they go through various phases of SDLC. A variety of tasks and activities are carried out in each phase of software development, and these are expensive and vulnerable to errors. Furthermore, software development approaches are modular and are associated with multiple developers and multi-tasking teams from various locations and different time zones.[10,11] Hence, proper coordination, cooperation, and communication are highly required among the multiple developers and development teams for the timely completion of the software development, and it can be done using various kinds of software repositories.

A software repository is a central storage location that contains a large amount of information about the development of software systems.[16] The information inside the software repositories is stored in the form of software packages, and these packages are shared among different developers from various locations associated with the development of a software project. It also contains a wealth of valuable information about the evolutionary history and the activities of developers for different software projects. The information stored inside the software repositories is used by different stakeholders for tracking and managing the processes of software projects.[20,32] Thus, software repositories are one of the most important parts of today's software development process. Since a large amount of data is available in software repositories, an extensive research effort is needed for a better understanding of these repositories. Nowadays, software engineering researchers have started to enhance the development of software by mining and analyzing these software repositories. The area of software engineering that is used for analyzing and understanding software repositories and extracts useful information for making intelligent decisions is known as mining software repositories (MSR). MSR leverages data mining to address the real-world software engineering problems.[35,40] It is very useful for a better understanding of the development of software, for estimating the time required for developing software, and for planning the various evolutionary aspects of the software project.

Modern software systems are larger and more complex than ever before. During the development of these large and complex software systems, a massive number of software bugs are continuously generated, and it becomes the main reason for the delay in software completion. Software companies spend the majority of their time and cost repairing these software bugs. As a result, a bug repository is the main repository in SDLC, and it has a significant impact on the completion and timely delivery of software. A software bug repository consists of all the defects that are encountered during the testing of software systems.[15,21] All the details regarding software bugs are collected in the form of a bug report from all the developers of software development at various locations and stored in a bug repository. For better software bug management, the majority of software companies use bug tracking systems such as Bugzilla, Jira, Eclipse, and Gnats. The role of these bug tracking systems is to record bug details in the form of bug reports, and these bug reports can be effectively used by triage for fixing the software bugs in the stipulated time. The process of assigning appropriate developers for fixing the software bug is known as bug triaging.[43,44] Bug triaging is one of the tedious and difficult tasks in software development.

A huge amount of research is going on to analyze the software bug repositories such as bug triaging, classification of software bugs, priority and severity prediction of software bugs, and estimating the bug fixing time by individual developers. Although software repositories provide a large amount of information about the software, there is still plenty of research to be done. The information of bug reports is textual. The majority of the traditional techniques uses the information retrieval[36,29] and machine learning algorithm[23,27,42] for analyzing the software repositories. These techniques consider the software bug repository-related problems as simple text mining problems. As a result, the outputs obtained by using these techniques are crisp. But in the real world, these techniques are not best suited for analyzing the software bug repositories. As software development is a modular approach with multiple developers and a multi-tasking team, a particular software bug may belong to more than one module and more than one developer can be associated with this. Hence, a better approach is needed for analyzing these software bugs. Recently, Fuzzy logic techniques have emerged as one of the best techniques in the area of text mining. Fuzzy logic techniques can be effectively applied over the software bug repositories for triaging, classification, severity, and priority prediction and estimating the time required for fixing software bugs.

The rest of the chapter is as follows: In Section 5.2, various software repositories are described in detail. In Section 5.3, software bug repositories and different research areas in software bug repositories are illustrated. Section 5.4, fuzzy logic based techniques are introduced and fuzzy logic based techniques for software bug repositories are discussed with the help of a case study and finally, the future trends and conclusion for the software repositories in the forthcoming era are illustrated in Section 5.5.

5.2   SOFTWARE REPOSITORIES

Software repositories are the central storage location of various software developments where all the information regarding the processes of software development is stored.[30,31] It is maintained offline and online by different stakeholders associated with the software development from various locations.[1] During the lifetime of software development, the numbers of software repositories and their use are increasing rapidly as several users are extracting the data for their study and changing it as per their need in these repositories. All the repositories are the result of the day-to-day interaction among different stakeholders based on their cooperation, coordination, and communication and the evolutionary changes to various software artifacts associated with software development.[24,28] The various software repositories, their instances along examples are shown in Table 5.1.

5.2.1 HISTORICAL REPOSITORIES

Historical repositories store information about the evolution and progress of a software project over a long time such as source control repositories, software bug repositories, and archived communications. All the information related to the history of software development and the changes made by individual developers during the software maintenance are stored in the source control repositories.[6] The bug repositories consist of the bug report that is encountered during different phases of software development. The archived communication keeps all the day-to-day communication and discussion among various stakeholders (such as developer, development lead, and project manager) associated with the development of software.

5.2.2 RUN-TIME REPOSITORIES

Run-time repositories consist of detailed information on the execution and usage of various applications at different deployment (single or multiple) sites. The run-time repositories are well understood by the deployment logs which consist of all the activities that are carried out by a producer as well as consumer associated with a software project.[42] It records the information related to the execution and deployment of a software project.

5.2.3 CODE REPOSITORIES

Code repositories keep all the information related to the source code of software projects. It consists of hundreds and thousands of lines of source codes that are freely available.[8] These source codes can be easily searchable and downloadable by different third-party users for their needs.

TABLE 5.1   Different Software Repositories and Their Instances with Example.

Type of repository	Instances of repository	Example
Historical repository	Source control repository	Perforce, CVS, ClearCase, Subversion, etc.
	Software bug repository	Eclipse, Bugzilla, Net beans, Kaggle, etc.
	Achieved communication	Emails, mailing lists, instance messages, etc.
Run-time repository	Deployment logs	Amazon Web Service code deploy, TeamCity,
		CodeShip, Jenkins, etc.
Code repository	Collection of source code	GitHub, Google Code, Sourceforge.net, etc.

5.3   SOFTWARE BUG REPOSITORY

Software systems became larger and more complex over the last few decades as they go through different phases, from requirement analysis to maintenance phases. In each phase of software development, a huge number of tasks are being performed and a large number of software bugs are continuously generated. A software bug is a flaw or error that is detected during the testing of software by the tester or test engineers.[15] The software bugs are stored in the form of a bug report consisting of all the information related to a software bug such as bug id, platform, component, version, assignee (name of the developer to whom the bug is assigned), summary, description, priority, severity, and the current status of the software bug.[36,49] An example of a bug report is shown in Figure 5.1. The bug report belongs to the eclipse dataset with bug id 277877. It belongs to the product platform and IDE component. The name of the assignee is Roy Ganor to whom the bug is assigned. The current status of the bug is NEW and the importance of the bug is P3 normal. The summary of the bug "The value in 'Limit visible items per group' field is applied both for the Tasks and Problems view." The bug is reported on 26 May 2009 by Kalin and modified on 06 September 2019. The hardware belongs to Window XP PC.

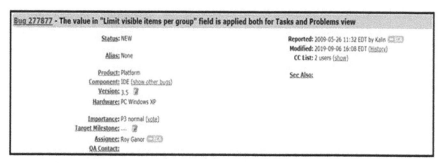

FIGURE 5.1 Bug report 277877 of eclipse datasets.
https://bugs.eclipse.org/bugs

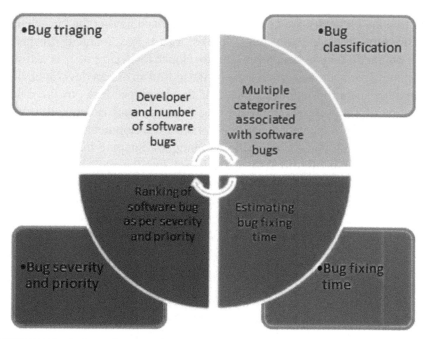

FIGURE 5.2 Software bug repository.

A software bug repository is the collection of a large number of such bug reports. Bug reports are continuously generated and reported during the software development and maintenance in the software bug repository.[21]

Some of the important bug repositories are Eclipse, Bugzilla, Net beans, and Kaggle, etc., which consist of detailed information about software bugs. The main areas of research in the software bug repository are bug triaging, bug classification, severity and priority prediction for software bugs, and prediction of bug fixing time. The software bug and its main area of research are shown in Figure 5.2.

5.3.1   BUG TRIAGING

A large number of bugs are continuously generated during the development of software, and finding appropriate developers to fix these software bugs is always a challenging task. Bug fixing is one of the important tasks in the software development process. The process of finding a suitable developer for fixing software bugs is known as bug triaging.[2,4] Initially, the bug triaging processes are done manually. A person having the knowledge and experience of software bugs, known as a triage, manually inspects each bug and assigns an appropriate developer for fixing it. It is a time-consuming and tedious work for the triage to find an expert developer, and sometimes inappropriate assignment of the developer leads to the delay in the bug fixing process.[13,14] If the bug fixing process is delayed, it will have a profound impact on the cost of software development. To overcome this problem, automatic bug triaging techniques emerged in recent years. Machine learning algorithms and artificial intelligence techniques are widely used for automatic bug triaging.[43,44,46] Although, automatic bug triaging techniques are available for finding an expert developer to fix the software bugs, still there exist many challenges that are unattained and need a significant amount of research.

5.3.2   BUG CLASSIFICATION

The modular approach of software development is incorporated multiple developers and a multi-tasking team. A software failure may be the result of multiple causes and may belong to multiple modules. Hence, the proper software bug classification techniques are the greatest necessity to ensure the timely fixing of software bugs. Software bug classification is a technique to categorize a software bug into different categories. [23]Classification of software bugs is done to get a deeper insight into the failure of a software bug. Existing classification algorithms use the terms present in a software bug and based on the terms, the software bugs are classified. The different types of classification techniques for software bugs are:

(1) Binary classification of software bug – In this case, the classification is done based on the two possible conditions (present or absent) of bug terms. It is used to classify software bugs into bugs or non-bugs categories.

(2) Multiclass classification of software bug – In this case, there are multiple categories, and a software bug belongs to only one category and does not belong to other categories at a particular time.

(3) Multi-label classification of software bug – In this case, there are multiple categories, and a software bug may belong to more than one category at a particular time.

Although a lot of machine learning-based algorithms had been developed in the last decade, still finding the root cause of the software bug is a challenging task in software development. Hence, the proper classification of software bugs is required to improve the quality of the software. This makes it an essential part of the software development.

5.3.3   BUG SEVERITY AND PRIORITY

The assessment of software bugs is done based on the severity and priority of the software bugs. The impact of the software bug on the functionality of the software is the severity of that software bug.[33,34] The software bugs that need immediate attention are identified and fixed as quickly as possible based on the severity level. Whereas the priority of the software bug is measured by the degree of risk involved in it. The severity of software bugs is characterized into various levels such as low (S4), minor (S3), major (S2), and critical (S1) to estimate the impact of the software bug.[3] For example, the low severity means the bug will not cause any significant breakdown to the system, and the critical severity means it has a profound impact on the system and the software bug must be fixed immediately. Similarly, the priority of the software bug is characterized in various levels such as P1, P2, P3, P4, and P5.[17,39] P1 represents the highest priority for the software bug, and it needs to be addressed as soon as possible. Although

P5 represents the lowest priority and is kept unfixed for a long period, the default priority is P3, and it is assigned by the bug triage if there is an uncertainty about the priority of the software bug. The developer then changes the priority based on their assessment of the software bug.

5.3.4 BUG FIXING TIME

In the modern-day software development, a large number of bugs are generated continuously for the large-scale software projects. Therefore, the project team needs to predict the time required for fixing such a large number of bugs while efficiently managing available resources. Bug fixing time refers to the total amount of time required to fix a bug starting from the time the bug is reported to the time it is fixed.[22] Bug fixing time is one of the most important aspects for improving the quality of the software as well as improving the software project management. Many factors affect the bug fixing time such as the source of the bug and bug type. Based on the bug type, the severity and priority for each software bug are assigned. Although severity and priority for the reported bugs are assigned, it is always challenging to estimate the bug fixing time as different bugs required different amounts of time for resolving it. A better estimation for the bug-fixing time will increase the customer satisfaction as well as improve the project planning.

The initial bug fixing is performed manually, which is very time-consuming and tedious. In the last decade, several researchers have started analyzing bug reports and estimating the bug-fixing time using machine learning algorithms. A lot of automatic bug-fixing time algorithms have been developed based on regression or classification techniques.[3,23,27] These machine learning algorithms extract the attributes of each bug report to estimate the bug-fixing time. Although automatic bug triaging techniques are available for estimating the bug-fixing time, still a lot of improvements are needed in these machine learning algorithms. All these machine learning algorithms treat software bugs in the same way as it works for plain text. But there are multiple reasons for a particular software bug, and these machine learning algorithms are not sufficient to deal with all the multiple reasons. Hence, better automatic algorithms are needed for better estimation of bug-fixing time.

5.4 COMPUTATIONAL INTELLIGENCE-BASED FUZZY LOGIC TECHNIQUE FOR SOFTWARE BUG REPOSITORY

In this section, the basic concepts of fuzzy logic, its advantages over a normal crisp set, and its application toward bug triaging are discussed. A software bug report in a software bug repository consists of the attributes such as bug id, assignee, product, component, summary, description, severity, priority, and current status of a software bug.[36,49] For a better understanding of the bug report, a sample bug report is shown in Figure 5.1, and it is taken from the Eclipse software bug datasets[a]. All the existing bug triaging techniques are focused on three different aspects of software bugs such as the most frequent terms, categories, and topics associated with software bugs.[19,29,41] In this chapter, the most frequent terms associated with software bugs are used for fuzzy modeling and analysis. Let us consider a set of the active developer (Assignee Name) and the number of bugs assigned to them (Bug Count) for the Eclipse software bug dataset within the range of 300–400 bug counts, and it is illustrated in Table 5.2. The Assignee Name column represents the name of the developer and the Bug Counts column represents the number of bugs assigned to the corresponding developer. *The names of the developers associated with the original Eclipse software bug dataset are not listed in this paper for privacy purposes.*

TABLE 5.2 Assignee Name and Their Bug Count for: A Sample from Eclipse Bug Dataset.

Assignee name	Bug count
AN-1	316
AN-2	325
AN-3	348
AN-4	393
AN-5	397

TABLE 5.3 A Sample of Assignee Name and Their Most Frequent Terms.

Assignee name	Most frequent terms
AN-1	INSTAL, UPDAT, FAIL, USE, REPOSITORI, PLUGIN, DIALOG, FILE, SITE, ERROR
AN-2	INSTAL, UPDAT, PUBLISH, SITE, SUPPORT, USE, ADD, AVAIL, DIALOG, PROFIL

TABLE 5.3   *(Continued)*

Assignee name	Most frequent terms
AN-3	COMPIL, CLASS, SEARCH, TYPE, WARN, FILE, METHOD, JAVA, ERROR, FOLDER
AN-4	TARGET, PLUGIN, EDITOR, LAUNCH, PLATFORM, PRODUCT, ADD, EXPERT, EXTENS, PDE
AN-5	TABL, TREE, COLUMN, ITEM, WINDOW, SELECT, CONTROL, LISTEN, USE, CHANG

[a]https://bugs.eclipse.org/bugs

The summary and assignee names from the sample bug data are extracted. Initially, pre-processing for the sample bug is performed. Then, the most frequent terms are generated for all the Assignee Names and a sample is shown in Table 5.3. Once the vocabulary list for each Assignee Name is generated, in the next step the AssigneeName-UniqueTerm matrix is created for all the Assignee Names, and it is shown in Table 5.4. For Assignee Name AN-1, the term count for term ADD is 8 and similarly, it is 17 for AN-2 and so on.

TABLE 5.4   A Sample of Assignee Name – Term Matrix.

Assignee name	ADD	AVAIL	BUILD	CHANG	COLUMN	CONFIGUR	CRASH
AN-1	8	13	2	8	3	4	0
AN-2	17	13	2	2	2	9	1
AN-3	12	2	16	5	0	1	3
AN-4	22	2	14	7	1	23	1
AN-5	17	0	0	16	34	0	22

Now it can be observed from Table 5.4 that one term is associated with more than one developer. When a new bug is entered in the bug repository, all the traditional machine-learning algorithm uses different similarity techniques such as Cosine, Jaccard, Dice, and SMTP similarity techniques for finding the expert developer. The main problem with all the existing machine learning algorithms is treating bug triaging as a simple text problem, and that leads to incorrect software bug allocation. These methods often classify a new bug to those developers who have a large number of bugs assigned and yet to fix or who have no experience of fixing this kind of bug. As a result, it delays the bug triaging process and becomes the main reason for the software reallocation. Thus these techniques are not sufficient for finding an expert developer.

In recent years, fuzzy logic based techniques have emerged and been widely used in various fields of text mining and data mining. The main advantage of fuzzy logic techniques over crisp logic is that it calculates the membership grade of individual developers associated with software bugs. According to Lotfi A Zadeh,[48] a fuzzy set A is a set of ordered pairs in the universe of discourse U, and it is defined as

$$A = \{(x, \mu_A(x)) \mid x \in U\} \tag{5.1}$$

where $\mu_A(x)$ denotes the degree of membership of x in A, and it ranges from 0 to 1 inclusively. Now using the definition of the fuzzy set (eq 5.1), the membership value of each Assignee Name and their associated terms can be calculated. Let us consider, A as the set of Assignee Name and T as the set of unique terms associated with A. The total number of Assignee Names is $n(A)$, and the total count of a unique term associated with all Assignee Names is $n(T)$. The fuzzy membership of each unique term $1 \leq j \leq n(T)$ with each Assignee Name $1 \leq i \leq n(A)$ is given by $\mu(A_i, T_j)$ and it is defined as

$$\mu(A_i, T_j) = \frac{A_i, T_j}{\sum_i^{n(A)} A_i, T_j}, \quad \forall 1 \leq j \leq n(T) \tag{5.2}$$

The fuzzy membership $\mu(A_i, T_j)$ values for a unique term of a particular Assignee Name is the ratio of the count of each term with the total count of that term associated with all Assignee Names. The fuzzy membership values for all Assignee Names in the training dataset are calculated using (eq 5.2) and shown in Table 5.5.

TABLE 5.5 Fuzzy Assignee Name – Term Matrix.

Assignee name	ADD	AVAIL	BUILD	CHANG	COLUMN	CONFIGUR	CRASH
AN-1	0.11	0.43	0.06	0.22	0.07	0.11	0.00
AN-2	0.22	0.43	0.06	0.05	0.05	0.24	0.04
AN-3	0.16	0.07	0.47	0.13	0.00	0.03	0.11

TABLE 5.5   *(Continued)*

Assignee name	ADD	AVAIL	BUILD	CHANG	COLUMN	CONFIGUR	CRASH
AN-4	0.29	0.07	0.41	0.18	0.02	0.62	0.04
AN-5	0.22	0.00	0.00	0.42	0.86	0.00	0.81

For Assignee Name AN-1, the fuzzy membership value for the term ADD is 0.11 and similarly, it is 0.22 for AN-2, and so on. Once the fuzzy membership values are calculated for all the Assignee Names, the next step is to find expert Assignee Names for a new bug. Let us consider two new bugs with their term count
NB1 = <1, 1, 0, 0, 0, 0, 0> and NB2 = <1, 0, 0, 0, 0, 0, 0>
For the new bug NB1, it having 1 term count for ADD and 1 term count for AVAIL; and for new bug NB2, it having only 1 term count for ADD. The fuzzy membership values will be calculated by dividing each term with the total term as discussed in (eq 5.2), and the corresponding values are

NB1 = <0.5, 0.5, 0, 0, 0, 0, 0> and NB2 = <1.0, 0, 0, 0, 0, 0, 0>

Now using the fuzzy similarity technique,[26] the similarity of new bugs with the Fuzzy Assignee Name – Term matrix is shown in Table 5.6. From the similarity values, it can be observed that NB1 can be fixed by AN-1 as well as AN-2 as they having the highest similarity values. Similarly, NB2 can be fixed by AN-1 as well as AN-2. Thus, by using fuzzy logic techniques a set of active Assignee Names can be predicted, and if one Assignee is busy in other work, he can still assist other Assignee who can fix the bug. Thus, it can solve the reallocation of the software bug and decrease the bug-fixing time for software bugs.

TABLE 5.6   Fuzzy Similarity of New Bugs with Fuzzy Assignee Name – Term Matrix.

Assignee name	NB1	NB2
AN-1	0.286	0.314
AN-2	0.203	0.225
AN-3	0.071	0.005
AN-4	0.119	0.026
AN-5	0.115	0.010

Although the normal fuzzy logic technique improves the bug triaging, still a lot of improvements are needed for effective bug triaging. People who report software bugs are unfamiliar with the technical terminology used in software development and are unable to write technically about the cause of the software bugs. As a result, software bugs are unlabeled, vague, and noisy. It becomes difficult for the triage to assign appropriate developers for fixing it. Furthermore, software development is a modular approach with a multi-skill development team, it is possible that a particular bug may occur due to more than one reason and at the same time, multiple developers can fix the bug. A normal fuzzy set is not sufficient to read the relationship between a developer and software bugs. Hence, Intuitionistic fuzzy sets (IFS) are more effective as compared to normal fuzzy sets as it deals with membership values, nonmembership values, and hesitancy values of developer and software bug relation. According to Atanassov,[5] an IFS A is defined as

$$A = \{(x, \mu_A(x), v_A(x)) \mid x \in U\} \tag{5.3}$$

where $\mu_A(x)$ denotes the degree of membership and $v_A(x)$ denotes the degree of non-membership of x in A. The ranges of both membership and non-membership are from 0 to 1 inclusively with the condition $0 \leq \mu_A(x) + v_A(x) \leq 1$. The hesitancy degree is represented by $\pi_A(x) = 1 - (\mu_A(x) - v_A(x))$ and $0 \leq \pi_A(x) \leq 1$. The IFS non-membership values $v_A(x)$ are calculated using the IFS membership values $\mu_A(x)$ and it is known as the IFS complement generator. Sugeno and Terano's complement generator[38] and Yager's complement generator[47] are two of the most commonly used IFS complement generators in the literature and are given as follows:

1. Sugeno and Terano's Complement Generator

$$v_A(x) = \frac{1 - \mu_A(x)}{1 + \lambda \mu_A(x)}; \lambda > 0 \tag{5.4}$$

2. Yager's Complement Generator

$$v_A(x) = \left(1 - \mu_A(x)^\lambda\right)^{1/\lambda}; \lambda > 0 \tag{5.5}$$

where λ is a constant. The IFS membership values are calculated using (eq 5.2) and it is shown in Table 5.7, the non-membership values are calculated using (eq 5.4) with $\lambda = 0.5$ and are shown in Table 5.8 and the hesitancy values are shown in Table 5.9, respectively. For Assignee Name AN-1, the IFS membership value for the term ADD is 0.105, the non-membership value is 0.850, and the hesitancy value is 0.045. Similarly, for AN-2, the membership value for the term ADD is 0.224, the non-membership value is 0.698, and the hesitancy value is 0.078, and so on.

TABLE 5.7   IFS Membership Values for Assignee Name – Term Matrix.

Assignee name	ADD	AVAIL	BUILD	CHANG	COLUMN	CONFIGUR	CRASH
AN-1	0.105	0.433	0.059	0.211	0.075	0.108	0.000
AN-2	0.224	0.433	0.059	0.053	0.050	0.243	0.037
AN-3	0.158	0.067	0.471	0.132	0.000	0.027	0.111
AN-4	0.289	0.067	0.412	0.184	0.025	0.622	0.037
AN-5	0.224	0.000	0.000	0.421	0.850	0.000	0.815

TABLE 5.8   IFS Non-Membership Values for Assignee Name – Term Matrix.

Assignee name	ADD	AVAIL	BUILD	CHANG	COLUMN	CONFIGUR	CRASH
AN-1	0.850	0.466	0.914	0.714	0.892	0.846	1.000
AN-2	0.698	0.466	0.914	0.923	0.927	0.675	0.946
AN-3	0.780	0.903	0.428	0.814	1.000	0.960	0.842
AN-4	0.621	0.903	0.488	0.747	0.963	0.288	0.946
AN-5	0.698	1.000	1.000	0.478	0.105	1.000	0.131

TABLE 5.9   IFS Hesitancy Value for Assignee Name – Term Matrix.

Assignee name	ADD	AVAIL	BUILD	CHANG	COLUMN	CONFIGUR	CRASH
AN-1	0.045	0.101	0.027	0.075	0.033	0.046	0.000
AN-2	0.078	0.101	0.027	0.024	0.023	0.082	0.017
AN-3	0.062	0.030	0.101	0.054	0.000	0.013	0.047
AN-4	0.090	0.030	0.100	0.069	0.012	0.090	0.017
AN-5	0.078	0.000	0.000	0.101	0.045	0.000	0.054

Once the IFS values of all the Assignee are computed for all the Assignee Names, the next step is to find expert Assignee Names for a new bug. For the new bug NB1 and NB2, the IFS membership values, non-membership values, and hesitancy values are calculated.

For NB1: IFS membership values <0.5, 0.5, 0.0, 0.0, 0.0, 0.0, 0.0>

IFS on-membership values <0.4, 0.4, 1.0, 1.0, 1.0, 1.0, 1.0>

IFS hesitancy values <0.1, 0.1, 0.0, 0.0, 0.0, 0.0, 0.0>

For NB2: IFS membership values <1.0, 0.0, 0.0, 0.0, 0.0, 0.0, 0.0>

IFS non-membership values <0.0, 1.0, 1.0, 1.0, 1.0, 1.0, 1.0>

IFS hesitancy values <0.0, 0.0, 0.0, 0.0, 0.0, 0.0, 0.0>

TABLE 5.10 IFS Similarities of New Bug NB1 with Fuzzy Assignee Name – Term Matrix.

Assignee name	SIMY[45]	SIMS[37]	SIMCCL[9]	SIMG[12]	SIMHM[25]	SIMJ[18]
AN-1	0.946	0.893	0.201	0.431	0.823	0.836
AN-2	0.921	0.878	0.222	0.339	0.805	0.818
AN-3	0.821	0.827	0.456	0.396	0.765	0.774
AN-4	0.806	0.789	0.049	0.135	0.689	0.704
AN-5	0.788	0.797	0.496	0.345	0.735	0.743

The available IFS similarity techniques[9,12,18,25,37,45] can be effectively used for finding the similarity of the new bugs NB1 and NB2 with all the Assignee Names. The IFS similarity of NB1 with IFS Assigne Name – Term matrix (IFS membership, non-membership, and hesitancy membership) is shown in Table 5.10.

TABLE 5.11 IFS Similarities of New Bug NB2 with Fuzzy Assignee Name – Term Matrix.

Assignee name	SIMY[45]	SIMS[37]	SIMCCL[9]	SIMG[12]	SIMHM[25]	SIMJ[18]
AN-1	0.948	0.899	0.171	0.442	0.825	0.838
AN-2	0.924	0.878	0.211	0.429	0.813	0.825
AN-3	0.836	0.827	0.436	0.412	0.776	0.784
AN-4	0.822	0.789	0.072	0.183	0.708	0.722
AN-5	0.807	0.797	0.504	0.381	0.752	0.760

The IFS similarity measures for NB2 are shown in Table 5.11. It can be observed that different IFS similarity techniques predict different Assignee Names for fixing the new bugs. User-defined threshold value α can also be used in the form of fuzzy α-cut over the IFS similarity to predict a set of Assignee Names for fixing the new bug. Thus, by using IFS techniques, a set of active Assignee Names can be predicted for fixing a new bug. Fuzzy logic techniques and IFS techniques can also be effectively applied in the area of software bug classification, prediction of severity and priority of software bugs, and estimating the time for bug fixing of software bug repository. The algorithm for bug triaging using fuzzy logic based techniques and IFS-based techniques are implemented using R-Programming[b] and the results are shown in the section.

[b]https://www.r-project.org

5.5   CONCLUSION

Over the last few decades, researchers and practitioners have started analyzing the software repositories to enhance and improve the software development process. Mining of software helped to some extend for the better understanding of the software repositories and the process involved in the development of software. Still, this area is wide open and a lot of research yet to be done. Softwares are developing rapidly in present era. As a result, there is a huge demand for high-quality software. The development of efficient software with the stipulated time and minimum cost is the main goal of all software organizations.

In this chapter, the problem faced by software industries and how these problems can be solved using fuzzy logic techniques are being discussed. This chapter enlightens the different areas of software bugs and their improvement using fuzzy logic based techniques. Software development is a modular approach and multiple developers are associated with it from various locations. Fuzzy logic techniques are best suited for analyzing these software bug repositories. It plays an important role in the better understanding of different processes associated with software bug repositories. Furthermore, the use of fuzzy logic techniques will improve bug triaging, classification of software bugs, and fixing the software bugs on time. The challenges and problems faced by software organizations are solved to great extent using fuzzy logic based on computational intelligence techniques.

In future, advanced fuzzy logic techniques can be applied over software bug repositories such as interval-valued fuzzy logic systems, hesitant fuzzy systems, and type-2 fuzzy systems for the better understanding of different processes associated with software bug repositories. Furthermore, fuzzy logic approaches can be applied to other software repositories to enhance and improve performance. These can also be used for understanding the cross-repository dependencies in software development.

KEYWORDS

- **computational intelligent techniques**
- **software repositories**
- **software bug repository**
- **mining software repositories**
- **fuzzy logic**
- **bug triaging**
- **software engineering**

REFERENCES

1. Abdellatif, A.; Badran, K.; Shihab, E. Msrbot: Using Bots to Answer Questions from Software Repositories. *Empirical Softw. Eng.* **2020,** *25* (3), 1834–1863.
2. Alazzam, I.; Aleroud, A.; Al Latifah, Z.; Karabatis, G. Automatic Bug Triage in Software Systems Using Graph Neighborhood Relations for Feature Augmentation. *IEEE Trans. Comput. Soc. Syst.* **2020,** *7* (5), 1288–1303.
3. Almhana, R.; Ferreira, T.; Kessentini, M.; Sharma, T. Understanding and Characterizing Changes in Bugs Priority: The Practitioners' Perceptive. In *2020 IEEE 20th International Working Conference on Source Code Analysis and Manipulation (SCAM)*; IEEE, 2020; pp 87–97.
4. Almhana, R.; Kessentini, M. Considering Dependencies between Bug Reports to Improve Bugs Triage. *Automated Softw. Eng.* **2021,** *28* (1), 1–26.
5. Atanassov, K. T. Intuitionistic Fuzzy Sets. In *Intuitionistic Fuzzy Sets*; Physica: Heidelberg, 1999; pp 1–137.
6. D'Ambros, M.; Gall, H.; Lanza, M.; Pinzger, M. Analysing Software Repositories to Understand Software Evolution. In *Software Evolution*; Springer: Berlin, Heidelberg, 2008; pp 37–67.

7. Bagnato, A.; Barmpis, K.; Bessis, N.; Cabrera-Diego, L A.; Di Rocco, J., Di Ruscio, D.; Gergely, T. et al. Developer-Centric Knowledge Mining from Large Open-Source Software Repositories (CROSSMINER). In *Federation of International Conferences on Software Technologies: Applications and Foundations*; Springer: Cham, 2017; pp 375–384.

8. Chaturvedi, K. K.; Sing, V. B.; Singh, P. Tools in Mining Software Repositories. In *2013 13th International Conference on Computational Science and Its Applications*; IEEE, 2013; pp 89–98.

9. Chen, S-M.; Cheng, S-H.; Lan, T-C. A Novel Similarity Measure between Intuitionistic Fuzzy Sets Based on the Centroid Points of Transformed Fuzzy Numbers with Applications to Pattern Recognition. *Inf. Sci.* **2016,** *343,* 15–40.

10. Chen, T-H.; Thomas, S. W.; Hassan, A. E. A Survey on the Use of Topic Models When Mining Software Repositories. *Empirical Softw. Eng.* **2016,** *21* (5), 1843–1919.

11. Destro, G. A.; de França, B. B. N. Mining Software Repositories for the Characterization of Continuous Integration and Delivery. In *Proceedings of the 34th Brazilian Symposium on Software Engineering*, 2020; pp. 664–669.

12. Garg, H.; Kumar, K. Distance Measures for Connection Number Sets Based on Set Pair Analysis and Its Applications to Decision-Making Process. *Appl. Intell.* **2018,** *48* (10), 3346–3359.

13. Ge, X.; Zheng, S.; Wang, J.; Li, H. High-Dimensional Hybrid Data Reduction for Effective Bug Triage. *Math. Probl. Eng.* **2020,** *2020.*

14. Guo, S.; Zhang, X.; Yang, X.; Chen, R.; Guo, C.; Li, H.; Li, T. Developer Activity Motivated Bug Triaging: Via Convolutional Neural Network. *Neural Process. Lett.* **2020,** *51* (3), 2589–2606.

15. Hamdy, Abeer, and Gloria Ezzat. Deep mining of Open Source Software Bug Repositories. *Int. J. Comput. App.* **2020,** 1-9.

16. Hassan, A. E. The Road Ahead for Mining Software Repositories. In *2008 Frontiers of Software Maintenance*; IEEE, 2008; pp 48–57.

17. Iqbal, S.; Naseem, R.; Jan, S.; Alshmrany, S.; Yasar, M.; Ali, A. Determining Bug Prioritization Using Feature Reduction and Clustering with Classification. *IEEE Access* **2020,** *8,* 215661–215678.

18. Jiang, Q.; Jin, X.; Lee, S-J.; Yao, S. A New Similarity/Distance Measure between Intuitionistic Fuzzy Sets Based on the Transformed Isosceles Triangles and Its Applications to Pattern Recognition. *Expert Syst. App.* **2019,** *116,* 439–453.

19. Jonsson, L.; Borg, M.; Broman, D.; Sandahl, K.; Eldh, S.; Runeson, P. Automated Bug Assignment: Ensemble-Based Machine Learning in Large Scale Industrial Contexts. *Empirical Softw. Eng.* **2016,** *21* (4), 1533–1578.

20. Kagdi, H.; Collard, M. L.; Maletic, J. I. A Survey and Taxonomy of Approaches for Mining Software Repositories in the Context of Software Evolution. *J. Softw. Maintenance Evol. Res. Pract.* **2007,** *19* (2), 77–131.

21. Kanwal, J.; Maqbool, O. Managing Open Bug Repositories through Bug Report Prioritization Using SVMs. In *Proceedings of the International Conference on Open-Source Systems and Technologies*, Lahore, Pakistan, 2010; pp 22–24.

22. Kashiwa, Y.; Ohira, M. A Release-Aware Bug Triaging Method Considering Developers' Bug-Fixing Loads. *IEICE Trans. Info. Syst.* **2020,** *103* (2), 348–362.

23. Kumaresh, S.; Baskaran, R. Mining Software Repositories for Defect Categorization. *J. Commun. Softw. Syst.* **2015**, *11* (1), 31–36.

24. Linstead, E.; Bajracharya, S.; Ngo, T.; Rigor, P.; Lopes, C.; Baldi, P. Sourcerer: Mining and Searching Internet-Scale Software Repositories. *Data Min. Knowl. Discov.* **2009**, *18* (2), 300–336.

25. Ngan, R. T.; Cuong, B. C.; Ali, M. H-Max Distance Measure of Intuitionistic Fuzzy Sets in Decision Making. *Appl. Soft Comput.* **2018**, *69*, 393–425.

26. Panda, R. R.; Nagwani, N. K. Software Bug Categorization Technique Based on Fuzzy Similarity. In *2019 IEEE 9th International Conference on Advanced Computing (IACC)*; IEEE, 2019; pp 1–6.

27. Pandey, N.; Sanyal, D. K.; Hudait, A.; Sen, A. Automated Classification of Software Issue Reports Using Machine Learning Techniques: An Empirical Study. *Innov. Syst. Softw. Eng.* **2017**, *13* (4), 279–297.

28. Papamichail, M.; Diamantopoulos, T.; Chrysovergis, I.; Samlidis, P.; Symeonidis, A. User-Perceived Reusability Estimation Based on Analysis of Software Repositories. In *2018 IEEE Workshop on Machine Learning Techniques for Software Quality Evaluation (MaLTeSQuE)*; IEEE, 2018; pp 49–54.

29. Peng, X.; Zhou, P.; Liu, J.; Chen, X. Improving Bug Triage with Relevant Search. In *SEKE*, 2017; pp 123–128.

30. Poncin, W.; Serebrenik, A.; Van Den Brand, M. Process Mining Software Repositories. In *2011 15th European Conference on Software Maintenance and Reengineering*; IEEE, 2011; pp 5–14.

31. Rao, D. V.; Sarma, V. V. S. A Computational Intelligence Approach to Software Component Repository Management. In *Innovations in Intelligent Machines-5*; Springer: Berlin, Heidelberg, 2014; pp. 109–132.

32. Reza, S. M.; Badreddin, O.; Rahad, K. Modelmine: A Tool to Facilitate Mining Models from Open Source Repositories. In *Proceedings of the 23rd ACM/IEEE International Conference on Model Driven Engineering Languages and Systems: Companion Proceedings*, 2020; pp 1–5.

33. Serban, C.; Vescan, A. Predicting Reliability by Severity and Priority of Defects. In *Proceedings of the 2nd ACM SIGSOFT International Workshop on Software Qualities and Their Dependencies*, 2019; pp 27–34.

34. Sharma, G.; Sharma, S.; Gujral, S. A Novel Way of Assessing Software Bug Severity Using Dictionary of Critical Terms. *Procedia Comput. Sci.* **2015**, *70*, 632–639.

35. Siddiqui, T.; Ahmad, A. Data Mining Tools and Techniques for Mining Software Repositories: A Systematic Review. *Big Data Analy.* **2018**, 717–726.

36. Soltani, M.; Hermans, F.; Bäck, T. The Significance of Bug Report Elements. *Empirical Softw. Eng.* **2020**, *25* (6), 5255–5294.

37. Song, Y.; Wang, X.; Lei, L.; Xue, A. A New Similarity Measure between Intuitionistic Fuzzy Sets and Its Application to Pattern Recognition. In *Abstract and Applied Analysis*, Vol. 2014; Hindawi, 2014.

38. Sugeno, M.; Terano, T. A Model of Learning Based on Fuzzy Information. *Kybernetes*, 1977.

39. Tan, Y.; Xu, S.; Wang, Z.; Zhang, T.; Xu, Z.; Luo, X. Bug Severity Prediction Using Question-and-Answer Pairs from Stack Overflow. *J. Syst. Softw.* **2020**, *165*, 110567.

40. Thomas, S. W.; Hassan, A. E.; Blostein, D. Mining Unstructured Software Repositories. In *Evolving Software Systems*; Springer: Berlin, Heidelberg, 2014; pp 139–162.

41. Tran, H. M.; Le, S. T.; Van Nguyen, S.; Ho, P. T. An Analysis of Software Bug Reports Using Machine Learning Techniques. *SN Comput. Sci.* **2020,** *1* (1), 1–11.

42. Tutko, A. Designing an Effective User Interface for Analyzing Software Repositories. In *2020 IEEE Symposium on Visual Languages and Human-Centric Computing (VL/HCC)*; IEEE, 2020; pp 1–2.

43. Xi, S-Q.;, Yao, Y.; Xiao, X-S.; Xu, F.; Lv, J. Bug Triaging Based on Tossing Sequence Modeling. *J. Comput. Sci. Technol.* **2019,** *34* (5), 942–956.

44. Xuan, J.; Jiang, H.; Hu, Y.; Ren, Z.; Zou, W.; Luo, Z.; Wu, X. Towards Effective Bug Triage with Software Data Reduction Techniques. *IEEE Trans. Knowl. Data Eng.* **2014,** *27* (1), 264–280.

45. Ye, J. Cosine Similarity Measures for Intuitionistic Fuzzy Sets and Their Applications. *Math. Comput. Model.* **2011,** *53* (1–2), 91–97.

46. Yadav, A.; Singh, S.; Suri, S. Ranking of Software Developers Based on Expertise Score for Bug Triaging. *Info. Softw. Technol.* **2019,** *112*, 1–17.

47. Yager, R. R. On the Measure of Fuzziness and Negation Part I: Membership in the Unit Interval, 1979, 221–229.

48. Zadeh, L. A. On Fuzzy Algorithms. In *Fuzzy Sets, Fuzzy Logic, and Fuzzy Systems: Selected Papers*; Zadeh, L. A., Ed.; 1996; pp. 127–147.

49. Zhou, C.; Li, B.; Sun, X.; Guo, H. Recognizing Software Bug-Specific Named Entity in Software Bug Repository. In *2018 IEEE/ACM 26th International Conference on Program Comprehension (ICPC)*; IEEE, 2018; pp. 108–10811.

CHAPTER 6

Software Measurements from Machine Learning to Deep Learning

SOMYA GOYAL*

Manipal University Jaipur, Jaipur, Rajasthan, India

*E-mail: somyagoyal1988@gmail.com

ABSTRACT

Software measurement (SM) is an umbrella activity during the entire software development cycle. Measurements and metrics of the attributes are indispensable for successful completion of project and effective delivery of software product. This chapter discusses SMs using deep learning (DL) techniques from the perspective of an empirical study. It is evident that an inaccurate prediction or estimation during the software development processes leads to loss of money and loss of projects. Since the beginning of software engineering, a wide range of methods are being deployed for measuring the software attributes. At present, the conventional techniques are not so apt for SMs due to excessive complex attributes of very large software. Machine learning (ML) has been the answer to all market needs in the past 30 years. It is noticed that ML is quite good to perform measurements in software engineering processes, but it is not the best method and needs enhancements. DL is the extension to ML, which is now being extensively used for SMs. The chapter begins with an introduction to ML and DL techniques and their applications in SMs empirically. Then, it highlights the literature work carried out in the field of empirical SMs using DL techniques. One of the most important DL techniques is

Computational Intelligence Applications for Software Engineering Problems. Parma Nand, PhD, Rakesh Nitin, PhD, Arun Prakash Agrawal, PhD & Vishal Jain, PhD (Eds.)
© 2023 Apple Academic Press, Inc. Co-published with CRC Press (Taylor & Francis)

convolutional neural networks which is discussed as a case study. This study provides a practical orientation to the readers about the implementation of DL technique to SMs.

6.1   INTRODUCTION

An accurate defect prediction is essential so that the developer can deliver a good quality product within the resource limits. An early prediction of faulty modules helps to dedicate the testing efforts to those modules and saves the testing efforts. It ensures the high quality of the end-product. It is evident that[7] only few of overall ongoing projects are completed successfully for the cause of inaccurate measurements.[10] Accurate estimations are highly desirable in software engineering in empirical aspects. Software defect prediction (SDP) is an important activity which is being carried out in the software industry since the beginning. From the 1980s, machine learning (ML) started finding heavy applications in SDP. SDP is the prediction of faulty modules in advance to commence the testing phase. For this purpose, the metrics and knowledge from the past projects were used to train the classifiers and after that, the trained classifier is used to classify the modules as buggy or clean for unforeseen projects. In this way, SDP is a learning problem that is solved with ML techniques beautifully.[17] The process of SDP is depicted in Figure 6.1. The training of ML model is being done with past data, and testing is to be done with current projects, and the project modules are divided into two categories—"defective" and "nondefective." Hence, SDP is shown as two-class classification problem and solved using ML techniques and deep learning (DL) techniques.

In this study, the transitional shift of software measurements (SMs) from the usage of ML techniques to the usage of DL techniques is conveyed. ML is being used in the domain of SMs since the past 30 years, now it is transforming itself as DL to meet the requirements of changing needs of the industry. This work highlights the benefits and appropriateness of DL techniques in changing era of software development processes. Beginning from the introduction to ML and DL, this study turns the orientation to the survey of DL techniques that are being deployed, and then inferences are drawn from the case study discussed. This chapter brings empirical approach to SMs with DL techniques.

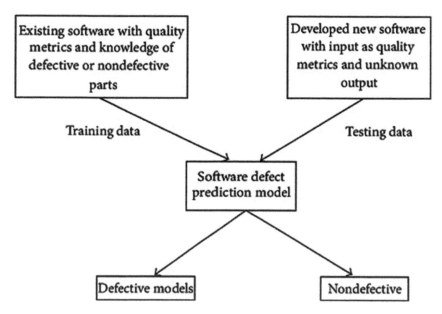

FIGURE 6.1   SDP process.

6.2   FROM MACHINE LEARNING TO DEEP LEARNING TRANSITION IN EMPIRICAL SOFTWARE MEASUREMENTS

This section takes a glimpse of ML and/or DL techniques available for applications in the software industry. A wide range of ML methods and DL techniques is available to find the optimal approach to SMs.[39] Techniques are categorized as, supervised learning techniques and unsupervised learning techniques,[13,62] as shown in Figures 6.2 and 6.3. Supervised techniques have labeled dataset and learn the patterns in presence of a supervisor, like classification and regression. Unsupervised methods do not have any trainers and they learn to segregate the nonlabeled data objects into groups on the basis of some similarity metrics.

DL[16] can be thought of as an extension to ML with better accuracy and capability to solve problems in a more realistic way. ML can be applied to the problem if the data are available and features are extracted by some domain experts[19–21] ML cannot learn features on their own selves. Here DL finds the reason of being popular nowadays as DL is capable to extract and learn features automatically without any human expert, as shown in Figure 6.4.

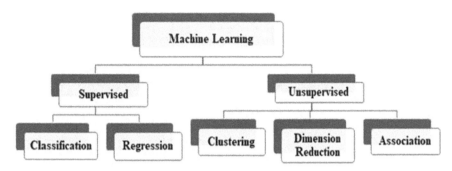

FIGURE 6.2   Machine learning techniques.

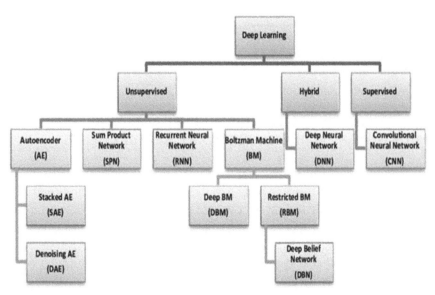

FIGURE 6.3   Deep learning techniques.

In the field of SMs, DL finds appropriate application as it seeks the semantics of software programs and makes the appropriate estimation from them.[63] It uses a deep representation structure to get better accuracy at estimation and prediction. It automatically extracts the features and results with better prediction power classifiers or predictors.[64] In these days, the research paradigm is shifting from ML to DL techniques. In the upcoming section, let us have a look at this shifting paradigm.

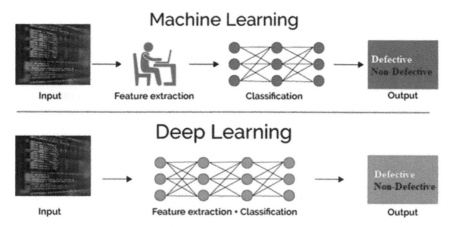

FIGURE 6.4 From machine learning to deep learning techniques.

6.3 CURRENT TRENDS OF USING ML AND DL IN SM

This section brings a torch on the studies from literature in relevance to the application of ML and DL in the domain of SMs (summarized in Table 6.1). The purpose of this section is to assess the popularity of ML in SM and to identify the gaps that require DL techniques over ML techniques. ML is a complementary method to conventional methods for SMs and improves the accuracy of measurements made while software development. It is observed that ML/DL techniques are intensively used for effort estimation, quality enhancement, change prediction, SCM activities, and many more processes falling under the vast field of software engineering.

TABLE 6.1 ML/DL State-of-Art-Of-The-Practice.

Study	Measurement Contributed	Technique used
[64]	Defect prediction	Deep learning
[46]	Effort and duration	SVM, MLP-ANN, and GLM
[27]	Effort	Heterogeneous ensembles
[14]	Enhancement effort	Support vector regression

TABLE 6.1 *(Continued)*

Study	Measurement Contributed	Technique used
[58]	Defect prediction	Deep learning
[65]	Defect prediction	Ensemble learning
[3]	Development effort	Hybrid MLPs
[40]	Development effort	Satin bower-bird optimizer
[30]	Development effort	RA, CART, SVR, RBF, kNN
[44]	Enhancement requests	BN, SVM, logistic regression
[37]	Defect prediction	GRARs-ANN
[29]	Defect prediction	ANNs
[57]	Quality prediction	TLBO-ANN
[63]	Defect prediction	Deep learning
[34]	Defect prediction	Convolutional neural network
[42]	Effort prediction	Genetic algorithm
[67]	Defect prediction	LR, BN, RBF,MLP, ADT
[5]	Development effort	DABE
[31]	Fault prediction	LSSVM
[2]	Defect prediction	Naïve Bayes
[41]	Fault-proneness	PSO-GA
[6]	Fault-proneness	ANN,NB, SVM, DT
[42]	Fault-proneness	Genetic algorithm
[48]	Fault-proneness	Ensemble learning
[50]	Fault-Proneness	SVM
[52]	Fault-proneness	Naïve Bayes
[53]	Effort prediction	Tree-based learning
[4]	Effort prediction	ANN
[8]	SBSE	SBSE
[28]	Missing value	kNN
[48]	Fault prediction	Ensemble learning
[66]	Defect prediction	Feature selection
[68]	Defect prediction	Tree-based learning
[36]	Defect prediction	Deep learning
[55]	Fault prediction	Parameter optimization
[60]	Test effort estimation	MLP-ANN
[61]	Fault prediction	ANN-SOM

TABLE 6.1   *(Continued)*

Study	Measurement Contributed	Technique used
[59]	Fault prediction	Binary + (GA/PSO/ACO)
[17]	Fault prediction	ANN, SVM, NN, Trees
[23]	Fault prediction	ANN, SVM, NN, Trees

6.4   DEEP LEARNING MODEL: EXPERIMENTAL SET-UP

The DL model is deployed in SMs basically in two situations—(1) when there is need of automation for feature selection, the steps are given in Figure 6.5, and (2) when there is necessity to assess the semantics of software coding to detect the faulty modules. The steps to build the DL models are shown in Figure 6.6.

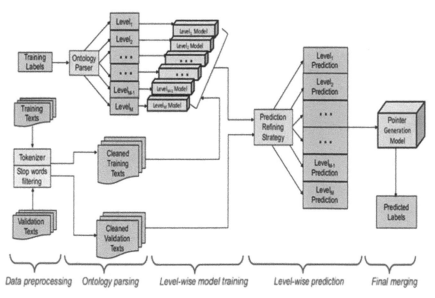

FIGURE 6.5   Steps to build deep learning model to assess semantics of software code.

As shown in Figure 6.5, the deep layers of model perform semantic analysis steps at multiple phases and bring prediction with better accuracy. In previous works, only static code metrics, OO metrics were being used for prediction task. Now, with deep structures, the semantics of the code can be checked for fault proneness. It allows the both levels—syntactic level and semantic level aspects to be considered while judging the software module as defective or nondefective. DL has introduced more power to the prediction and estimation tools in software industry. This is the reason that the trend is shifting to DL from ML nowadays.

Another scenario with DL structures is given in Figure 6.6. It is usecase for convolution neural networks. It is shown that how the features from the input software can be learnt automatically by the model. At the dense and deep layers, features are extracted and refined. Ultimately, the features are translated from primitive to high-level. In this way, the features are learnt, model is trained, and then these are used for prediction in case of unforeseen future projects.

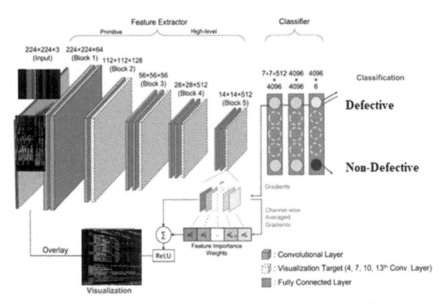

FIGURE 6.6　Steps to build DL model to select features automatically.

6.5　PERFORMANCE CRITERIA FOR MODEL EVALUATION

- The performance of prediction models is necessary to be evaluated to confirm the usefulness of the model. In literature, there are many criteria, which are used in customized combination to assess the performance of predictors[26,33]:

- *Confusion matrix*: It is a two-dimensional matrix (see in Fig. 6.7) that represents the information about the correctly classified and wrongly classified data points by the candidate prediction model.

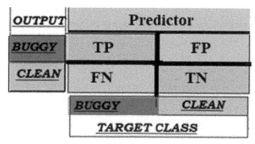

FIGURE 6.7   Confusion matrix.

- #Sensitivity (or #Recall) = # count of TP / (#count of TP + #count of FN).
- #Specificity = #count of TN / (#count of TN + #count of FP).
- #Precision = #count of TP / (#count of TP + #count of FP).
- #F1-Score = 2 * (#Recall * #Precision) / (#Recall + #Precision).
- The area covered under the ROC is computed and the closer it is found to the unit area, the model is assumed to be more accurate.[26]
- #Accuracy = (#count of TP + #count of TN) / (#count of TP + #count of FP + #count of TN + #count of FN).
- The above criteria are selected based on the popularity in the literature.[9,11,15,35]

Next, a case study is being discussed to have a clearer view of SMs using DL.

6.6 CASE STUDY

This section discusses a case study to demonstrate the implementation of DL techniques for SMs. The case study deploys long short-term memory (LSTM) for prediction software defects (see Fig. 6.8).[1] The measurement of software quality is a highly complex task and equally essential one for the effective completion of software project. In this case study, quality prediction is formulated as a classification task. And, for the sake simplicity, we have taken it as two-class classification problem, in which the software modules are being classified as "defective" or "nondefective" modules. The "defective" modules are the modules that are risky and prone to errors. The "nondefective" modules are those modules that are not-so-risky in the final product. The aim is to predict the faulty modules in early development phases so that testing efforts can be targeted to those faulty modules and the quality of product can be improved.

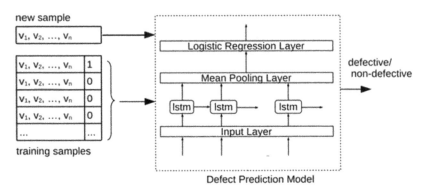

FIGURE 6.8   Experimental set-up.

LSTM model is being developed over PROMISE datasets. First up, features are extracted from the selected defect dataset. Next, the LSTM model is trained and tested. Finally, the performance is measured.

6.6.1   DATA DESCRIPTION

The case study has been carried out with defect dataset available in PROMISE repository.[18,45,54,56]

6.6.2   LAYERS AND IMPLEMENTATION

All SDP model is developed using standard implementation of LSTM in python. In the model, fully connected input layer, LSTM layer, means pooling layer.

6.6.3   PERFORMANCE CRITERIA

The criteria used for performance evaluation are precision, recall, F1-measure, and accuracy SDP classifier. Table 6.2 reports the obtained values observed in this experiment.

TABLE 6.2   Performance Measure.

Performance measure	Value observed
Recall	0.6286
F1-Score	0.5878
Accuracy	0.8215
Precision	0.543

The case study is designed in alignment with the current state-of-the-art of using DL for software quality prediction.[34] Since the past 30 years, ML has intensively been used in SDP field.[1,6,12,25,32,38,49,69] Now, the paradigm is shifting from ML to DL for better prediction power and for the expansion of the research in SM domain.

6.7   FUTURE SCOPE AND CONCLUSION

The shift in the paradigm from ML approach to DL is empirically conveyed in this chapter. Since the last three decades, many researchers have contributed to the SDP domain of SMs. Now, the paradigm has shifted to DL models, representations, and structures. This study brings empirically the application of DL methods in SMs. The literature review is carried out throughout the span of SDP field, that is, 1988–2021. The research trends reflect that the most popular ML technique is neural network which is now extended as its variants—deep neural network and convolution neural networks. A practical demonstration is given with a case study taking an example of LSTM. Here, I conclude the work with special references to DL[16] as a tool to perform SMs.

ML techniques are always found to be suitable for the measurements of software development activities and processes. Now, it is transforming itself into its variant—DL. DL is an extension to ML techniques with some more additional capabilities that make it more apt for the changing needs of the software industry.

KEYWORDS

- **software measurements**
- **defect prediction**
- **effort estimation**
- **machine learning**
- **deep learning**
- **neural networks**

REFERENCES

1. Abaei, G.; Selamat, A.; Fujita, H. An Empirical Study Based on Semi-Supervised Hybrid Self-Organizing Map for Software Fault Prediction. *Knowl. Based Syst.* **2015,** *74,* 28–39.

2. Arar, O. F.; Ayan, K. A Feature Dependent Naive Bayes Approach and Its Application to the Software Defect Prediction Problem. *Appl. Soft. Comput.* **2017,** *59,* 197–209.

3. Araujo, A.; Oliveira, L. I.; Meira, S. A Class of Hybrid Multilayer Perceptrons for Software Development Effort Estimation Problems. *J. Expert Syst. App.* **2017,** *90 C,* 1–12. 10.1016/j.eswa.2017.07.050

4. Azzeh, M.; Nassif, A. B.; Banitaa, S. Comparative Analysis of Soft Computing Techniques for Predicting Software Effort Based Use Case Points. *IET Softw.* **2018,** *12* (1), 19–29.

5. Benala, T. R.; Mall, R. DABE: Differential Evolution in Analogy-Based Software Development Effort Estimation. *Swarm. Evol. Comput.* **2018,** *38,* 158–172.

6. Boucher, A.; Badri, M. Software Metrics Thresholds Calculation Techniques to Predict Fault-Proneness: An Empirical Comparison. *Inf. Softw. Technol.* **2018,** *96,* 38–67.

7. Chaos Report. The Standish Group, 2015.

8. Chen, J.; Nair, V.; Menzies, T. Beyond Evolutionary Algorithms for Search-Based Software Engineering. *Inf. Softw. Technol.* **2017.** http://dx.doi.org/10.1016/j.infsof.2017.08.007

9. Czibula, G.; Marian, Z.; Czibula, I. G. Software Defect Prediction Using Relational Association Rule Mining. *Inf. Sci.* **2014,** *264,* 260–278.

10. Demarco, T. Controlling Software Projects: Management, Measurement and Estimation, 1982.

11. Dejaeger, K.; Verbraken, T.; Baesen, B. Toward Comprehensible Software Fault Prediction Models Using Bayesian Network Classifiers. *IEEE Trans. Softw. Eng.* **2013,** *39* (2), 237–257.

12. Erturk, E.; Sezer, E. A. A Comparison of Some Soft Computing Methods for Software Fault Prediction. *Expert Syst. App.* **2015,** *42,* 1872–1879.

13. Ethem, A. *Introduction to Machine Learning (Adaptive Computation and Machine Learning Series)*; MIT Press, 2014.

14. García-Floriano, A.; López-Martín, C.; Yáñez-Márquez, C.; Abran, A. Support Vector Regression for Predicting Software Enhancement Effort. *Inf. Softw. Technol.* **2018,** *97,* 99–109.

15. Gondra, I. Applying Machine Learning to Software Fault-Proneness Prediction. *J. Syst. Softw.* **2008,** *81* (5), 186–195.

16. Goodfellow, I.; Bengio, Y.; Courville, A. *Deep Learning*; MIT Press, 2016.

17. Goyal, S. Heterogeneous Stacked Ensemble Classifier for Software Defect Prediction. In *2020 Sixth International Conference on Parallel, Distributed and Grid Computing (PDGC)*; Waknaghat, Solan, India, 2020; pp 126–130. doi: 10.1109/PDGC50313.2020.9315754.

18. Goyal, S.; Parashar, A. Machine Learning Application to Improve COCOMO Model Using Neural Networks. *Int. J. Inform. Technol. Comput. Sci.* **2018,** *10* (3), 35–51. 10.5815/ijitcs.2018.03.05.

19. Goyal, S.; Bhatia, P. K. A Non-Linear Technique for Effective Software Effort Estimation Using Multi-Layer Perceptrons. In *International Conference on Machine Learning, Big Data, Cloud and Parallel Computing* (ComITCon 2019); IEEE, 2019a. 978-1-7281-0212-2

20. Goyal, S.; Bhatia, P. K. Feature Selection Technique for Effective Software Effort Estimation Using Multi-Layer Perceptrons. In *Proceedings of ICETIT 2019, Emerging Trends in Information Technology*, 2019b. 10.1007/978-3-030-30577-2_15

21. Goyal, S.; Bhatia, P. K. GA Based Dimensionality Reduction for Effective Software Effort Estimation Using ANN. *Adv. App. Math. Sci.* June **2019,** *18* (8), 637–649.

22. Goyal, S.; Bhatia, P. K. Comparison of Machine Learning Techniques for Software Quality Prediction. *Int. J. Knowl. Syst. Sci.* **2020,** *11* (2), 21–40. DOI: 10.4018/IJKSS.2020040102

23. Goyal S.; Bhatia, P. K. Software Quality Prediction Using Machine Learning Techniques. In *Innovations in Computational Intelligence and Computer Vision*: *Advances in Intelligent Systems and* Computing; Sharma, M. K., Dhaka, V. S., Perumal, T., Dey, N., Tavares, J. M. R. S., Eds.; , Vol. 1189Springer: Singapore, 2021; pp 551–560. https://doi.org/10.1007/978-981-15-6067-5_62

24. Goyal, S.; Bhatia, P. K. Empirical Software Measurements with Machine Learning. In *Computational Intelligence Techniques and Their Applications to Software Engineering Problems*; Bansal, A., Jain, A., Jain, S., Jain, V., Choudhary, A., Ed.; CRC Press: Boca Raton, 2021b; pp 49–64. https://doi.org/10.1201/9781003079996

25. Gyimothy, T.; Ferenc, R.; Siket, I. Empirical Validation of Object-Oriented Metrics on Open Source Software for Fault Prediction. *IEEE Trans. Softw. Eng.* **2005,** *31,* 897–910.

26. Hanley, J.; McNeil, B. J. The Meaning and Use of the Area under a Receiver Operating Characteristic ROC Curve. *Radiology* **1982,** *143,* 29–36.

27. Hosni, M.; Idri, A.; Abran, A. Investigating Heterogeneous Ensembles with Filter Feature Selection for Software Effort Estimation. In *Proceedings of the 27th International Workshop on Software Measurement and 12th International Conference on Software Process and Product Measurement on- IWSM Mensura'17*, 2017; pp 207–220.

28. Huang, J.; Keung, J. W.; Sarro, F. et al. Cross-Validation Based K Nearest Neighbor Imputation for Software Quality Datasets: An Empirical Study. *J. Syst. Softw.* **2017**. 10.1016/j.jss.2017.07.012

29. Jayanthi, R.; Florence, L. Software Defect Prediction Techniques Using Metrics Based on Neural Network Classifier. *J. Cluster Comput.* **2018,** 1–12.

30. Jodpimai, P.; Sophatsathit, P.; Lursinsap, C. Re-estimating Software Effort Using Prior Phase Efforts and Data Mining Techniques. *Innov. Syst. Softw. Eng.* **2018**.

31. Kumar, L.; Sripada, S. K.; Rath, S. K. Effective Fault Prediction Model Developed Using Least Square Support Vector Machine (LSSVM). *J. Syst. Softw.* **2018,** *137*, 686–712.

32. Laradji, I. H.; Alshayeb, M.; Ghouti, L. Software Defect Prediction Using Ensemble Learning on Selected Features. *Inf. Softw. Technol.* **2015,** *58*, 388–402.

33. Lehmann, E. L.; Romano, J. P. *Testing Statistical Hypothesis: Springer Texts in Statistics*; Springer: New York, 3rd ed., 2005. Corr. 2nd printing 2008 edition (Sept 10, 2008). ISBN-13: 978-0387988641.

34. Li, J.; He, P.; Zhu, J.; Lyu, M. Software Defect Prediction via Convolutional Neural Network. In *IEEE International Conference on Software Quality, Reliability and Security*, 2017; pp 318–328.

35. Malhotra, R. Comparative Analysis of Statistical and Machine Learning Methods for Predicting Faulty Odules. *Appl. Soft Comput.* **2014,** *21*, 286–297.

36. Manjula, C.; Florence, L. Deep Neural Network Based Hybrid Approach for Software Defect Prediction Using Software Metrics. *Cluster Comput.* **2018**. 10.1007/s10586-018-1696-z

37. Miholca, D.; Czibula, G.; Czibula, I. G. A Novel Approach for Software Defect Prediction through Hybridizing Gradual Relational Association Rules with Artificial Neural Networks. *Info. Sci. Info. Comput. Sci. Intell. Syst. App.: Int. J.* **2018,** *441* (C), 152–170.

38. Mishra, B.; Shukla, K. Defect Prediction for Object Oriented Software Using Support Vector Based Fuzzy Classification Model. *Int. J. Comput. Appl.* **2012,** *60*.

39. Mitchell, T. *Machine Learning*; McGraw-Hill, 1997.

40. Moosavi, S.; Bardsiri, V. Satin Bowerbird Optimizer: A New Optimization Algorithm to Optimize ANFIS for Software Development Effort Estimation. *Eng. App. Artif. Intell.* **2017,** *60*, 1–15.

41. Moussa, R.; Azar, D. A PSO-GA Approach Targeting Fault-Prone Software Modules. *J. Syst. Softw.* **2017,** *132*, 41–49.

42. Murillo-Morera, J.; Castro-Herrera, C.; Arroyo, J. An Automated Defect Prediction Framework using Genetic Algorithms: A Validation of Empirical Studies. *Intel. Artif.* **2016,** *19* (57), 114–137. 10.4114/ia.v18i56.1159.

43. Murillo-Morera, J.; Quesada-López, C.; Castro-Herrera, C. A genetic algorithm based framework for software effort prediction. *J. Softw. Eng. Res. Dev.* **2017,** *5*, 1–4.

44. Nizamani, Z. A.; Liu, H.; Niu, Z. Automatic Approval Prediction for Software Enhancement Requests. *J. Autom. Softw. Eng.* **2018,** *25* (2), 347–381.

45. PROMISE. http://promise.site.uottawa.ca/SERepository.

46. Pospieszny, P.; Czarnacka-Chrobot, B.; Kobylinski, A. An Effective Approach for Software Project Effort and duration Estimation with Machine Learning Algorithms. *J. Syst. Softw.* **2018,** *137,* 184–196.

47. Rathore, S. S.; Kumar, S. Linear and Non-Linear Heterogeneous Ensemble Methods to Predict the Number of Faults in Software Systems. *Knowl. Based Syst.* **2017,** *119,* 232–256.

48. Rathore, S. S.; Kumar, S. Towards an Ensemble Based System for Predicting the Number of Software Faults. *Expert Syst. App.* **2017,** *82,* 357–382.

49. Rodríguez, D.; Ruiz, R.; Riquelme, J. C. A Study of Subgroup Discovery Approaches for Defect Prediction. *Inf. Softw. Technol.* **2013,** *55* (10), 1810–1822.

50. Rong, X.; Li, F.; Cui, Z. A Model for Software Defect Prediction Using Support Vector Machine Based on CBA. *Int. J. Intell. Syst. Technol. App.* **2016,** *15* (1), 19–34.

51. Ross, S. M. *Probability and Statistics for Engineers and Scientists*, 3rd ed.; Elsevier Press, 2005. ISBN:81-8147-730-8.

52. Ryu, D.; Baik, J. Effective Multi-Objective Naïve Bayes Learning for Cross-Project Defect Prediction. *J. Appl. Soft Comput.* **2016,** *49* (C), 1062–1077.

53. Satapathy, S.; Achary, B.; Rath, S. Early Stage Software Effort Estimation Using Random Forest Technique Based on Use Case Points. *IET Softw. Inst. Eng. Technol.* **2016,** *10* (1), 10–17.

54. Sayyad, S.; Menzies, T. The PROMISE Repository of Software Engineering Databases; University of Ottawa: Canada, 2005. http://promise.site.uottawa.ca/ SERepository.

55. Tantithamthavorn, C.; Mcintosh, S.; Hassan, A. An Empirical Comparison of Model Validation Techniques for Defect Prediction Models. *IEEE Trans. Softw. Eng.* **2017,** *43* (1), 1–18.

56. Thomas, J. McCabe, a complexity measure. *IEEE Trans. Softw. Eng.* **1976,** *2* (4), 308–320.

57. Tomar, P.; Mishra, R.; Sheoran, K. Prediction of Quality Using ANN Based on Teaching-Learning Optimization in Component-Based Software Systems. *J. Softw. Pract. Exper.* **2018,** 1–15.

58. Tong, H.; Liu, B.; Wang, S. Software Defect Prediction Using Stacked Denoising Autoencoders and Two-Stage Ensemble Learning. *Inf. Softw. Technol.* **2018,** *96,* 94–111.

59. Turabieh, H.; Mafarja, M.; Li, X. Iterated Feature Selection Algorithms with Layered Recurrent Neural Network for Software Fault Prediction. *Expert Syst. App.* **2019,** *122,* 27–42. https://doi.org/10.1016/j.eswa.2018.12.033.

60. Vig, V.; Kaur, A. Test Effort Estimation and Prediction of Traditional and Rapid Release Models Using Machine Learning Algorithms. *J. Intell. Fuzzy Syst.* **2018,** *35,* 1657–1669. 10.3233/JIFS-169703.

61. Viji, C.; Rajkumar, N.; Duraisamy, S. Prediction of Software Fault-Prone Classes Using an Unsupervised Hybrid SOM Algorithm. *Cluster Comput.* **2018.** https://doi. org/10.1007/s10586-018-1923-7.

62. Witten, L. H.; Frank, E.; Hell, M. A. *Data Mining: Practical Machine Learning Tools and Techniques*, 3rd edn. ACM Sigsoft Software Engineering Notes. Morgan Kaufmann: Burlington, 2011; pp 90–99.

63. Wang, S.; Liu, T.; Nam, J.; Tan, L. Deep Semantic Feature Learning for Software Defect Prediction. *IEEE Trans. Softw. Eng.* **2018**.

64. Yang, X.; Lo, D.; Xia, X.; Zhang, Y.; Sun, J. Deep Learning for Just-in-Time Defect Prediction. In *Proceedings of the IEEE International Conference on Software Quality, Reliability and Security*, 2015.

65. Yang, X.; Lo, D.; Xin, X.; Sun, J. TLEL: A Two-Layer Ensemble Learning Approach for Just-in-time Defect Prediction. *J. Inf. Softw. Technol.* **2017**, *87*, 206–220.

66. Yu, Q.; Jiang, S.; Zhang, Y. A Feature Matching and Transfer Approach for Cross-Company Defect Prediction. *J. Syst. Softw.* **2017**, 132, 366–378. 10.1016/j.jss.2017.06.070.

67. Zhang, Y.; Lo, D.; Xia, X.; Sun, J. Combined Classifier for Cross-Project Defect Prediction: An Extended Empirical Study. *Front. Comput. Sci.* **2018**, *12* (2), 280–296.

68. Zhou, L.; Li, R.; Zhang, S.; Hang, H. Imbalanced Data Processing Model for Software Defect Prediction. *Wireless Pers. Commun.* **2017**, *6*, 1–14. https://doi.org/10.1007/s11277-017-5117-z.

69. Zhou, Y.; Leung, H. Empirical Analysis of Object-Oriented Design Metrics for Predicting High and Low Severity Faults. *IEEE Trans. Softw. Eng.* **2006**, *32*, 771–789.

CHAPTER 7

Time Series Forecasting Using ARIMA Models: A Systematic Literature Review of 2000s

VIDHI VIG*

Department of Computer Science, S.G.T.B. Khalsa College, University of Delhi, North Campus, New Delhi 110007, India

E-mail: vidhi.ipu@gmail.com

ABSTRACT

Software engineering involves gathering and designing requirements followed by coding and maintenance. With growing advancements, machine learning and forecasting have essentially become an extremely important part of the software process. Time series forecasting is one such area where researchers have shown consistent interests since the early 1900s. The current paper aims to survey and explore the research conducted in time series forecasting particularly related to ARIMA (AutoRegressive Integrated Moving Average) models. For this, two decades of research have been mined and studied to understand the trend and efficiency of this model in time series forecasting.

7.1 INTRODUCTION

Future is unpredictable and advancements in computer science have made this unpredictability predictable. Machine learning techniques, artificial

Computational Intelligence Applications for Software Engineering Problems. Parma Nand, PhD, Rakesh Nitin, PhD, Arun Prakash Agrawal, PhD & Vishal Jain, PhD (Eds.)

intelligence, deep learning, etc., are a few of the super-achievers that have made it all possible. Time series analysis and forecasting are yet another procedure that is rigorously used in many real-world applications.[5,40,41,46]

Time series forecasting and analysis prudently collect and examine past observations and cautiously study these observations to propose efficient models that elaborate the integral framework of the series.[20] These models further produce results that may be used to make predictions. An efficient model can provide meaningful insights about the future problems. It has been observed that a good amount of effort and time is spent to develop models that fit the real-world problems perfectly and engender the least amount of errors. Consequently, a variety of models and studies have been conducted for this purpose.

Time series collected data can be classified either linear and nonlinear or volatile and nonvolatile. Models like support vector machine (SVM)[61] and artificial neural network (ANN) have been explored extensively with nonlinear time series data.[58] The linear categorization of data is extensively explored by AutoRegressive Integrated Moving Average (ARIMA).[27] While ANNs are popular for nonparametric analysis, ARIMA model has demonstrated its expertise with efficient and precise time series forecasting with linear data. However, both models have their pros and cons. The performance of ANNs in certain situations may be variable irrespective of its adaptability and flexibility.[34] ARIMA on the other hand cannot work with nonlinear data and the data to be stationary.[57] Therefore, it may not be wrong if it is concluded that none among them can be examined eclectically in all the circumstances.

This chapter presents an extensive summary of the state of the art of time series forecasting models and summarizes the evolution of models in the arena since early 2000.

7.2 METHOD

The current study is an extensive review of the literature and adheres to the original framework suggested by Kitchenham et al.[35] The main aim of the review is to gaze the review of literature, also known as secondary studies, of the researchers conducting their research in the area of time series analysis and forecasting. The literature review is further structured as follows.

7.2.1   RESEARCH QUESTIONS

The research questions explored by the current study are as follows:

RQ1: How many research papers have been published from the year 2000−2020?

RQ2: What were the key metrics used to determine the performance and accuracy of the ARIMA models?

RQ3: Are there any variations/extension to ARIMA models in time series forecasting?

RQ4: What were the key advantages of ARIMA over other time series forecasting methods?

RQ5: What were the key limitations of ARIMA over other time series forecasting methods?

RQ6: What were the sources of data in the selected research papers?

7.2.2   SEARCH PROCESS

This process was conducted manually for a variety of national and international conferences and journals starting from the year 2000 to 2021. The reason behind the selection of these journals and conferences was influenced by their published empirical studies or surveys in this area. These journals and conferences were extensively reviewed and published thereafter, ensuring the quality of the research.

7.2.3   INCLUSION AND EXCLUSION CRITERIA

- Articles extensively peer-reviewed and published during the years 2000–2020 were gathered and filtered for keywords like "time series forecasting," "prediction," "ARIMA."
- Survey and research papers including the above-mentioned keywords were also explored in details

Since the current study focuses on time series forecasting and analysis, only papers pertaining to ARIMA models were explored. Note, there may exist other models and customized techniques for time series forecasting, but the current study in order to be precise researched papers related to ARIMA only.

Papers and articles that were excluded from the study were as follows:

- Plain basic papers published without ISSN numbers.
- Papers defining survey strategies and procedures involving literature reviews in general.
- Vaguely and informally written review papers.
- Redundant reports of the same paper (papers published in the journal were considered over papers published in conferences by the same author in the same area).

7.2.4   QUALITY ASSESSMENT

Each paper selected was further investigated for the following quality criteria:

QC1: Inclusion and exclusion criteria proposed above have been achieved comprehensively by the study.

QC2: Length of the paper is not less than four pages.

QC3: Extensive literature review has been performed on the area of research.

QC4: Research is reproducible.

QC5: Number of citations.

7.2.5   DATA COLLECTION

The research papers were downloaded from Springer, IEEE, Elsevier, ACM, IEEE Xplore, Wiley, Taylor and Francis, Sage publications, Google, Google scholar, Dblp, arxiv, ResearchGate, etc.

7.2.6   DATA ANALYSIS

A deep and thorough investigation of each paper was made in order to see the inclusion and exclusion criteria. Papers meeting the inclusion criteria (given earlier) were then cited for their full length. Once gathered, they were studied thoroughly to answer the research questions. Duplicate papers and papers whose source was untraceable were removed. Further quality assessment (as mentioned earlier) was also made and papers thus obtained were analyzed. The results obtained are discussed in the section below.

7.3 RESULTS

This part of the paper presents a comprehensive summary of results of the study.

7.3.1 SEARCH RESULTS

The number of papers that were shortlisted in our literature review was 109. Papers were categorized for papers published from the year 2000 to 2020 particularly in the domain of forecasting time series using ARIMA models. The papers were then analyzed for their journals and conferences types, their publishing houses, authors, year, domain on which the time series forecasting was made, and variations or extensions to ARIMA modeling. A brief overview of the list of papers surveyed is given in Table 7.1. It can be observed from Figure 7.1 that Elsevier is the leading publishing house of time series forecasting using ARIMA models.

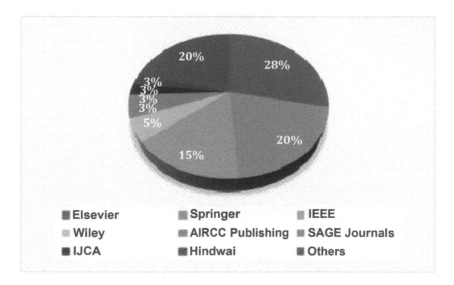

FIGURE 7.1   Publishing houses and the numbers of papers published since 2000.

TABLE 7.1 List of Papers Published in Time Series Forecasting Using ARIMA Models from 2000 to 2021.

Refs. no.	Year	Dataset	Period
[2]	2020	Economic data	2 years
[6]	2020	Purchase data	2015–2018
[11]	2020	COVID 19	2020
[13]	2020	Smoker's data	2006–2017
[60]	2020	Air quality data	1987–2015
[22]	2020	Climate data	1901–2002
[69]	2020	COVID 19	2020
[32]	2020	COVID 19	2020
[47]	2020	COVID 19	2020
[65]	2020	COVID 19	2020
[4]	2019	Solar radiation data	1981–2017
[80]	2019	Bitcoin dataset	2012–2016
[15]	2019	Economic data	2005–2008
[48]	2019	Coking coal prices	2005–2010
[3]	2019	Traffic	3 months
[24]	2018	Data of demand	2010–2015
[1]	2018	Pollution data	2013–2016
[67]	2018	Rainfall forecasting	1901–2002
[50]	2018	Bitcoin dataset	2013–2016
[66]	2018	Economic data	1985–2018
[54]	2018	Agricultural data	2001–2011
[82]	2018	China stock market	2016
[52]	2018	Meteorological data	2016–2018
[28]	2018	Disease data	9 years
[75]	2017	Tuberculosis data	2007–2016

TABLE 7.1 *(Continued)*

Refs. no.	Year	Dataset	Period
[77]	2017	Cyber data	2010–2012
[7]	2016	Ministry of Public Health	2005–2015
[49]	2016	New York City births (NYB)	1946–1959
[36]	2015	Smart meter readings	74 weeks
[44]	2016	Self-generated data	–
[25]	2016	Gold price data	2003–2014
[38]	2015	Comex copper spot price	2002–2014
[76]	2015	Annual runoff data	1952–2003
[59]	2015	Colombo Stock Exchange	2010–2013
[63]	2014	EUR/USD, EUR/GBP and EUR/CHF	1999–2012
[51]	2014	April 2012–February 2014	2012–2014
[31]	2014	Stock index	2001–2006
[10]	2014	Sunspots, electricity price and Indian stock data.	2010
[12]	2014	Stock index	2013
[37]	2014	Agricultural data	2014–2017
[19]	2013	Household electric power consumption	2006–2010
[21]	2013	Stock index	2007–2011
[62]	2013	Disease data	4 years
[39]	2012	Distance data	2001–2002
[42]	2012	NASDAQ, TAIEX, DJI	2006–2010
[34]	2011	Exchange rate data	2000–2004
[30]	2011	Electricity data	10 years
[23]	2010	Water data	1996–2004

TABLE 7.1   *(Continued)*

Refs. no.	Year	Dataset	Period
[26]	2010	Agricultural data	1999–2006
[45]	2010	Bus travel data	1 year
[16]	2010	Wind data	1 month
[14]	2009	RMB exchange data	2001–2006
[18]	2008	Property crime data	50 weeks
[9]	2007	Tourist data	2003–2004
[53]	2006	Palm price data	3 years
[29]	2006	Agricultural data	2000–2002
[81]	2003	Economic data	1980–1993
[79]	2002	Per capita personal income dataset	1929–1999
[72]	2002	Machinery data from Taiwan	1991–1996
[43]	2002	Australian tourist data	1975–1984
[34]	2001	Population dataset	1900–1999

7.4 DISCUSSION

RQ1: How many research papers have been published from the year 2000–2020?

The research paper extensively surveys research papers published in the field of time series analysis for ARIMA models. Though the methodology goes back to the 1900s, the current work limited itself to the research published from the 2000s. At the initial step of the study, close to 135 papers were downloaded that followed our inclusion and exclusion criteria. These papers were then filtered for papers published with proper ISBN number. After this step, 73 papers discussing extensively ARIMA models were shortlisted for the survey.

It can be observed from Figure 7.2 that there has been a consistent rise in research in this area from the year 2000. It was very interesting to observe how research in this area has caught pace since the 1900s. One important information for a researcher could be the name of the journals readily publishing the research findings in this area. It was observed that the *International Journal of Forecasting, Journal of Time Series Analysis,* followed by *Malaria* journal were the publishers where we found our maximum articles. Figure 7.3 presents the top journals and conferences in this area.

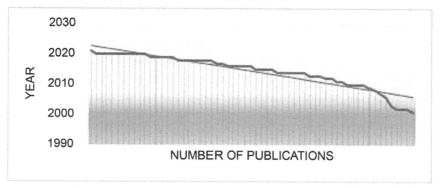

FIGURE 7.2   Number of publication (year wise).

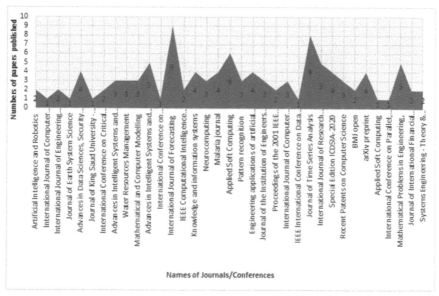

FIGURE 7.3   Number of publications (top journal/conference wise).

RQ2: What were the key metrics gauged to determine the performance and accuracy of the ARIMA models?

Root mean square error (RMSE) is one of the most frequently used identifier to measure the accuracy of the ARIMA models (Fig. 7.4). RMSE has been employed as a benchmark metric to gaze performance of the model in various quality assessments especially, air, meteorology, water, agriculture, and climate.[17] Studies cite mean absolute error (MAE) is also a popular metric for evaluation.[78] However, it has been observed that RMSE holds more importance for time-specific data. While MAE is desirable for uniformly distributed errors and model errors are more inclined be normally distributed rather than uniform distribution, RMSE is preferred over MAE. Many of the research papers produced a confusion matrix and analyzed the accuracy, precision, and recall of the proposed models too. However, choosing just one single metric describes only one side of the errors in the model, and thus emphasizes a particular aspect of the error. Many other studies observed variance, mean absolute percent error, mean square error, akaike information criterion,[74] mean absolute deviation (MAD) for analyzing errors and identification of best models.

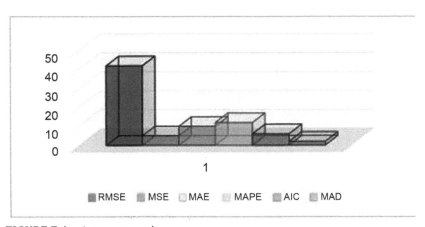

FIGURE 7.4   Accuracy metric.

RQ3: Are there any variations/extension to ARIMA models in time series forecasting?

ARIMA model was initially proposed by Box and Jenkins in 1976,[1] and since then it is referred to as Box–Jenkins models also.[37] This model analyzes the previous data of univariate time series to study time series pattern to predict future trends.

Though ARIMA has been most popularly used, but there exist papers that propose efficiency of variation of ARIMA models for forecasting. One such model is ARIMAX. It has been proposed that one can incorporate more than one-time series in the same model to forecast the range of other series, by tampering the function popularly known as "transfer function" (TF). This function, TF, can be employed to shape and predict the response series and to study effect of the intervention. The basic TF model used by ARIMA includes input variables to time series. Such a model is known as ARIMAX model by the studies[55] work referred ARIMAX model as dynamic regression. Studies observed that ARIMAX presented better results than ARIMA but the errors such as RMSE and MAE were both observed to be lower in ARIMA compared to ARIMAX by the studies.[56]

The seasonal time series ARIMA (SARIMA) model was primarily proposed by Box–Jenkins.[27] It is popularly explored to study and analyze various socioeconomic problems. This model was given magnificent results for short intervals, (50–100 observations). Moreover, SARIMA managed errors to handle digression among inputs and outputs. However, this model exploits the values that are not taken care of by measurement errors[8,68,70,71] proposed the fuzzy regression to decipher the fuzzy environment and reduce the modeling error[71] further suggested the fuzzy ARIMA (FARIMA) method involving fuzzy regression method to further fuzz the boundary of the ARIMA model. They found FARIMA better over SARIMA and ARIMA, as it presented the worst and best scenarios far more efficiently and demanded lesser number of observations than SARIMA.

RQ4: What were the key advantages of ARIMA over other time series forecasting methods?

Time series forecasting has been extensively explored for decades. The popularity of ARIMA models in this domain is widely known.[64] ARIMA being parametric methodology, works best with shorter cycles especially when the flexibility is barred due to insufficient availability of data. This methodology is cynic and unlike other time series forecasting methodologies, it does not require any know-how of socioeconomic relationships. Since ARIMA method need only the historical know-how of a time series to analyze and forecast results, it can improve the forecasting accuracy by restricting the number of parameters.

It has also been researched that in comparison forecasting methods such as SVM and ANN, ARIMA model is the most efficient in forecasting epidemics and natural calamities.[69]

RQ5: What were the key limitations of ARIMA over other time series forecasting methods?

ARIMA models are claimed to be "backward-looking" since they are very timid at forecasting the "turning-points."[36] However, this may not be a problem if the unseen turning points comeback to a long-run equilibrium.

Unlike other basic models, ARIMA models cannot update itself automatically on the availability of newer set of data.[23] Therefore, one must redo the entire modeling in such a scenario. Consequently, these models tend to be a little costlier when the scenarios are changing repeatedly and the number pf parameters are also higher.

It has also been observed that the skill and expertise of the researcher/ programmer influences the forecasted results due to subjectivity of the inputs it involves.[73]

RQ6: What were the sources of data in the selected research papers?

A good source of data for analysis has come from government open source sites. These sites are uploaded with data mainly for use of research only. Results obtained of such data are significantly reliable than other open-source data as they are clean and tidy (Fig. 7.5). Close to 10 papers, they were focusing on closed source datasets and a majority of them are centric to the organization, thereby making it both authentic and tidy. But such papers lack reproducibility as the data are not readily available to the researchers.

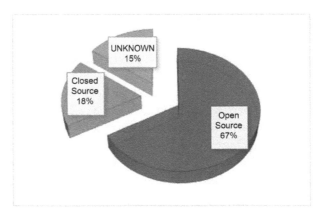

FIGURE 7.5   Source of data.

RQ7: What were the key areas where this forecasting has been implemented?

Time series forecasting has been extensively used in forecasting socio-economic problems. From daily and monthly average, global solar radiation, stocks, food companies, weather, air quality, power consumption, air pollution, draught, agriculture, rainfall, disease control, and measurement, water quality, data mining, crime investigation and monitoring, and commodity pricing and control. Broadly bifurcating these into social and economic categories (Fig. 7.6), it is observed that ARIMA model is very widely used in both the areas for time series forecasting.

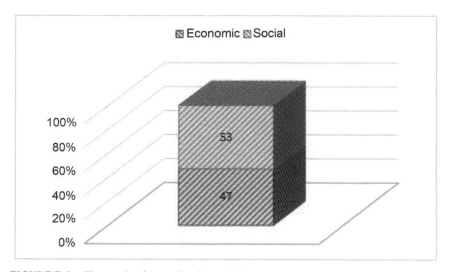

FIGURE 7.6   Time series forecasting in social/economic areas.

7.5 CONCLUSIONS

The current study presents a comprehensive literature review of time series forecasting using ARIMA models. The survey reviews more than 20 years of research to demonstrate a snapshot of the current trend in this area. The findings of the survey indicate that even though the time series forecasting using ARIMA models has a long and deep research history, there exists a consistent rise in the papers in this area. This further reflects the potential of research and growth for researchers in this domain.

KEYWORDS

- **time series**
- **forecasting**
- **prediction**
- **ARIMA**
- **modeling**

REFERENCES

1. Abhilash, M. S. K.; Thakur, A.; Gupta, D.; Sreevidya, B. Time Series Analysis of Air Pollution in Bengaluru Using ARIMA Model. In *Ambient Communications and Computer Systems*; Springer: Singapore, 2018; pp 413–426.

2. Aijaz, I.; Agarwal, P. A Study on Time Series Forecasting Using Hybridization of Time Series Models and Neural Networks. *Recent Adv. Comput. Sci. Commun.* **2020,** *13* (5), 827–832.

3. Alghamdi, T.; Elgazzar, K.; Bayoumi, M.; Sharaf, T.; Shah, S. Forecasting Traffic Congestion Using ARIMA Modeling. In *2019 15th International Wireless Communications & Mobile Computing Conference (IWCMC)*; IEEE, June 2019; pp 1227–1232.

4. Alsharif, M. H.; Younes, M. K.; Kim, J. Time Series ARIMA Model for Prediction of Daily and Monthly Average Global Solar Radiation: The Case Study of Seoul, South Korea. *Symmetry* **2019,** *11* (2), 240.

5. Aminikhanghahi, S.; Cook, D. J. A Survey of Methods for Time Series Change Point Detection. *Knowl. Inform. Syst.* **2017,** *51* (2), 339–367.

6. Anderson, P.; Llopis, E. J.; O'Donnell, A.; Manthey, J.; Rehm, J. Impact of Low and No Alcohol Beers on Purchases of Alcohol: Interrupted Time Series Analysis of British Household Shopping Data, 2015–2018. *BMJ Open* **2020,** *10* (10), e036371.

7. Anwar, M. Y.; Lewnard, J. A.; Parikh, S.; Pitzer, V. E. Time Series Analysis of Malaria in Afghanistan: Using ARIMA Models to Predict Future Trends in Incidence. *Malaria J.* **2016,** *15* (1), 1–10.

8. Asai, H. T. S. U. K.; Tanaka, S.; Uegima, K. Linear Regression Analysis with Fuzzy Model. *IEEE Trans. Syst. Man Cybern.* **1982,** *12*, 903–907.

9. Aslanargun, A.; Mammadov, M.; Yazici, B.; Yolacan, S. Comparison of ARIMA, Neural Networks and Hybrid Models in Time Series: Tourist Arrival Forecasting. *J. Statist. Comput. Simul.* **2007,** *77* (1), 29–53.

10. Babu, C. N.; Reddy, B. E. A Moving-Average Filter Based Hybrid ARIMA–ANN Model for Forecasting Time Series Data. *Appl. Soft Comput.* **2014,** *23*, 27–38.

11. Barman, A. Time Series Analysis and Forecasting of Covid-19 Cases Using LSTM and ARIMA Models. **2020.** *arXiv preprint arXiv:2006.13852.*

12. Banerjee, D. Forecasting of Indian Stock Market Using Time-Series ARIMA Model. In *2014 2nd International Conference on Business and Information Management (ICBIM)*; IEEE, Jan 2014; pp 131–135.

13. Beard, E.; West, R.; Michie, S.; Brown, J. Association of Prevalence of Electronic Cigarette Use with Smoking Cessation and Cigarette Consumption in England: A Time–Series Analysis between 2006 and 2017. *Addiction* **2020**, *115* (5), 961–974.

14. Bo, S. U. N.; Chi, X. I. E. RMB Exchange Rate Forecasting in the Context of the Financial Crisis. *Syst. Eng. Theory Practice* **2009**, *29* (12), 53–64.

15. Büyükşahin, Ü. Ç.; Ertekin, Ş. Improving Forecasting Accuracy of time Series Data Using a New ARIMA-ANN Hybrid Method and Empirical Mode Decomposition. *Neurocomputing* **2019**, *361*, 151–163.

16. Cadenas, E.; Rivera, W. Wind Speed Forecasting in Three Different Regions of Mexico, Using a Hybrid ARIMA–ANN Model. *Renew. Energy* **2010**, *35* (12), 2732–2738.

17. Chai, T.; Draxler, R. R. Root Mean Square Error (RMSE) or Mean Absolute Error (MAE)?–Arguments against Avoiding RMSE in the Literature. *Geosci. Model Develop.* **2014**, *7* (3), 1247–1250.

18. Chen, P.; Yuan, H.; Shu, X. Forecasting Crime Using the Arima Model. In *2008 Fifth International Conference on Fuzzy Systems and Knowledge Discovery*; IEEE, October 2008; Vol. 5, pp 627–630.

19. Chujai, P.; Kerdprasop, N.; Kerdprasop, K. Time Series Analysis of Household Electric Consumption with ARIMA and ARMA Models. In *Proceedings of the International Multi-Conference of Engineers and Computer Scientists*, March 2013; Vol. 1, pp 295–300.

20. De Gooijer, J. G.; Hyndman, R. J. 25 years of Time Series Forecasting. *Int. J. Forecast.* **2006**, *22* (3), 443–473.

21. Devi, B. U.; Sundar, D.; Alli, P. An Effective Time Series Analysis for Stock Trend Prediction Using ARIMA Model for Nifty Midcap-50. *Int. J. Data Mining Knowl. Manage. Process* **2013**, *3* (1), 65.

22. Dimri, T.; Ahmad, S.; Sharif, M. Time Series Analysis of Climate Variables Using Seasonal ARIMA Approach. *J. Earth Syst. Sci.* **2020**, *129* (1), 1–16.

23. Faruk, D. Ö. A Hybrid Neural Network and ARIMA Model for Water Quality Time Series Prediction. *Eng. App. Artif. Intell.* **2010**, *23* (4), 586–594.

24. Fattah, J.; Ezzine, L.; Aman, Z.; El Moussami, H.; Lachhab, A. Forecasting of Demand Using ARIMA Model. *Int. J. Eng. Busi. Manage.* **2018**, *10*, 1847979018808673.

25. Guha, B.; Bandyopadhyay, G. Gold Price Forecasting Using ARIMA Model. *J. Adv. Manage. Sci.* **2016**, *4* (2).

26. Han, P.; Wang, P. X.; Zhang, S. Y. Drought Forecasting Based on the Remote Sensing Data Using ARIMA Models. *Math. Comp. Modell.* **2010**, *51* (11–12), 1398–1403.

27. Harvey, A. C.; Todd, P. H. J. Forecasting Economic Time Series with Structural and Box-Jenkins Models: A Case Study. *J. Busi. Econ. Statist.* **1983**, *1* (4), 299–307.

28. He, Z.; Tao, H. Epidemiology and ARIMA Model of Positive-Rate of Influenza Viruses among Children in Wuhan, China: A Nine-Year Retrospective Study. *Int. J. Infect. Dis.* **2018**, *74*, 61–70.

29. Hossain, Z.; Samad, Q. A.; Ali, Z. ARIMA Model and Forecasting with Three Types of Pulse Prices in Bangladesh: A Case Study. *Int. J. Soc. Econ.* **2006**.

30. Jakaša, T.; Andročec, I.; Sprčić, P. Electricity Price Forecasting—ARIMA Model Approach. In *2011 8th International Conference on the European Energy Market (EEM)*; IEEE, May 2011; pp 222–225.
31. Junior, P. R.; Salomon, F. L. R.; de Oliveira Pamplona, E. ARIMA: An Applied Time Series Forecasting Model for the Bovespa Stock Index. *Appl. Mathem.* **2014**, *5* (21), 3383.
32. Kakar, A.; Nundy, S. COVID-19 in India. *J. R. Soc. Med.* **2020**, *113* (6), 232–233.
33. Kalpakis, K.; Gada, D.; Puttagunta, V. Distance Measures for Effective Clustering of ARIMA Time-Series. In *Proceedings 2001 IEEE International Conference on Data Mining*; IEEE, November 2001; pp 273–280.
34. Khashei, M.; Bijari, M. A Novel Hybridization of Artificial Neural Networks and ARIMA Models for Time Series Forecasting. *Appl. Soft Comput.* **2011**, *11* (2), 2664–2675.
35. Kitchenham, B.; Brereton, O. P.; Budgen, D.; Turner, M.; Bailey, J.; Linkman, S. Systematic Literature Reviews in Software Engineering–A Systematic Literature Review. *Info. Softw. Technol.* **2009**, *51* (1), 7–15.
36. Krishna, V. B.; Iyer, R. K.; Sanders, W. H. ARIMA-based Modeling and Validation of Consumption Readings in Power Grids. In *International Conference on Critical Information Infrastructures Security*; Springer: Cham, October 2015; pp 199–210.
37. Kumar, M.; Anand, M. An Application of Time Series ARIMA Forecasting Model for Predicting Sugarcane Production in India. *Studies Busi. Econ.* **2014**, *9* (1), 81–94.
38. Lasheras, F. S.; de Cos Juez, F. J.; Sánchez, A. S.; Krzemień, A.; Fernández, P. R. Forecasting the COMEX Copper Spot Price by Means of Neural Networks and ARIMA Models. *Resour. Policy* **2015**, *45*, 37–43.
39. Li, C.; Hu, J. W. A New ARIMA-based Neuro-Fuzzy Approach and Swarm Intelligence for Time Series Forecasting. *Eng. App. Artif. Intell.* **2012**, *25* (2), 295–308.
40. Liao, T. W. Clustering of Time Series Data—a Survey. *Pattern Recogn.* **2005**, *38* (11), 1857–1874.
41. Lee, C. M.; Ko, C. N. Short-term Load Forecasting Using Lifting Scheme and ARIMA Models. *Expert Syst. App.* **2011**, *38* (5), 5902–5911.
42. Li, C.; Chiang, T. W. Complex Neurofuzzy ARIMA Forecasting—A New Approach Using Complex Fuzzy Sets. *IEEE Trans. Fuzzy Syst.* **2012**, *21* (3), 567–584.
43. Lim, C.; McAleer, M. Time Series Forecasts of International Travel Demand for Australia. *Tour. Manage.* **2002**, *23* (4), 389–396.
44. Liu, C.; Hoi, S. C.; Zhao, P.; Sun, J. Online Arima Algorithms for Time Series Prediction. In *Proceedings of the AAAI Conference on Artificial Intelligence*, February 2016; Vol. 30, No. 1.
45. Madzlan, N.; Ibrahim, K. ARIMA Models for Bus Travel Time Prediction, 2010.
46. Mahalakshmi, G.; Sridevi, S.; Rajaram, S. A Survey on Forecasting of Time Series Data. In *2016 International Conference on Computing Technologies and Intelligent Data Engineering (ICCTIDE'16)*; IEEE, January 2016; pp 1–8.
47. Marbaniang, S. P. Forecasting the Prevalence of COVID-19 in Maharashtra, Delhi, Kerala, and India Using an ARIMA Model, 2020.
48. Matyjaszek, M.; Fernández, P. R.; Krzemień, A.; Wodarski, K.; Valverde, G. F. Forecasting Coking Coal Prices by Means of ARIMA Models and Neural Networks, Considering the Transgenic Time Series Theory. *Resour. Policy* **2019**, *61*, 283–292.

49. Mehrmolaei, S.; Keyvanpour, M. R. Time Series Forecasting Using Improved ARIMA. In *2016 Artificial Intelligence and Robotics (IRANOPEN)*; IEEE, April 2016; pp 92–97.

50. McNally, S.; Roche, J.; Caton, S. Predicting the Price of Bitcoin Using Machine Learning. In *2018 26th Euromicro International Conference on Parallel, Distributed and Network-Based Processing (PDP)*; IEEE, March 2018; pp 339–343.

51. Mondal, P.; Shit, L.; Goswami, S. Study of Effectiveness of Time Series Modeling (ARIMA) in Forecasting Stock Prices. *Int. J. Comput. Sci. Eng. App.* **2014,** *4* (2), 13.

52. Murat, M.; Malinowska, I.; Gos, M.; Krzyszczak, J. Forecasting Daily Meteorological Time Series Using ARIMA and Regression Models. *Int. Agrophys.* **2018,** *32* (2).

53. Nochai, R.; Nochai, T. ARIMA Model for Forecasting Oil Palm Price. In *Proceedings of the 2nd IMT-GT Regional Conference on Mathematics, Statistics and Applications*, June 2006; pp 13–15.

54. Ohyver, M.; Pudjihastuti, H. Arima Model for Forecasting the Price of Medium Quality Rice to Anticipate Price Fluctuations. *Procedia Comput. Sci.* **2018,** *135*, 707–711.

55. Pankratz, A. *Forecasting with Dynamic Regression Models*, Vol. 935; John Wiley & Sons, 2012.

56. Peter, Ď.; Silvia, P. ARIMA vs. ARIMAX–which Approach is Better to Analyze and Forecast Macroeconomic Time Series. In *Proceedings of 30th International Conference Mathematical Methods in Economics*, September 2012; Vol. 2, pp 136–140.

57. Piccolo, D. A Distance Measure for Classifying ARIMA Models. *J. Time Ser. Analy.* **1990,** *11* (2), 153–164.

58. Rani, S.; Sikka, G. Recent Techniques of Clustering of Time Series Data: A Survey. *Int. J. Comp. App.* **2012,** *52* (15).

59. Rathnayaka, R. K. T.; Seneviratna, D. M. K. N.; Jianguo, W.; Arumawadu, H. I. A Hybrid Statistical Approach for Stock Market Forecasting Based on Artificial Neural Network and ARIMA Time Series Models. In *2015 International Conference on Behavioral, Economic and Socio-cultural Computing (BESC)*; IEEE, October 2015; pp 54–60.

60. Rekhi, J. K.; Nagrath, P.; Jain, R. Forecasting Air Quality of Delhi Using ARIMA Model. In *Advances in Data Sciences, Security and Applications*; Springer: Singapore, 2020; pp 315–325.

61. Sapankevych, N. I.; Sankar, R. Time Series Prediction Using Support Vector Machines: A Survey. *IEEE Comput. Intell. Mag.* **2009,** *4* (2), 24–38.

62. Sato, R. C. Disease Management with ARIMA Model in Time Series. *Einstein* **2013,** *11* (1), 128.

63. Sermpinis, G.; Stasinakis, C.; Dunis, C. Stochastic and Genetic Neural Network Combinations in Trading and Hybrid Time-Varying Leverage Effects. *J. Int. Financ. Markets Inst. Money* **2014,** *30*, 21–54.

64. Sezer, O. B.; Gudelek, M. U.; Ozbayoglu, A. M. Financial Time Series Forecasting with Deep Learning: A Systematic Literature Review: 2005–2019. *Appl. Soft Comput.* **2020,** *90*, 106181.

65. Sharma, V. K.; Nigam, U. Modeling and Forecasting of Covid-19 Growth Curve in India. *Trans. Indian Natl. Acad. Eng.* **2020,** *5* (4), 697–710.

66. Siami-Namini, S.; Namin, A. S. Forecasting Economics and Financial Time Series: ARIMA vs. LSTM. **2018**. *arXiv preprint arXiv:1803.06386*.
67. Swain, S.; Nandi, S.; Patel, P. Development of an ARIMA Model for Monthly Rainfall Forecasting over Khordha District, Odisha, India. In *Recent Findings in Intelligent Computing Techniques*; Springer: Singapore, 2018; pp 325–331.
68. Tanaka, H. Possibilistic Regression Analysis Based on Linear Programming. *Fuzzy Regression Analysis*. **1992**.
69. Tandon, H.; Ranjan, P.; Chakraborty, T.; Suhag, V. Coronavirus (COVID-19): ARIMA Based Time-Series Analysis to Forecast near Future. **2020**. *arXiv preprint arXiv:2004.07859*.
70. Tseng, F. M.; Tzeng, G. H.; Yu, H. C.; Yuan, B. J. Fuzzy ARIMA Model for Forecasting the Foreign Exchange Market. *Fuzzy Sets Syst.* **2001**, *118* (1), 9–19.
71. Tseng, F. M.; Tzeng, G. H. A Fuzzy Seasonal ARIMA Model for Forecasting. *Fuzzy Sets Syst.* **2002**, *126* (3), 367–376.
72. Tseng, F. M.; Yu, H. C.; Tzeng, G. H. Combining Neural Network Model with Seasonal Time Series ARIMA Model. *Technol. Forecast. Soc. Change* **2002**, *69* (1), 71–87.
73. Van Der Voort, M.; Dougherty, M.; Watson, S. Combining Kohonen Maps with ARIMA Time Series Models to Forecast Traffic Flow. *Transport. Res. Part C: Emerg. Technol.* **1996**, *4* (5), 307–318.
74. Vasileiadou, E.; Vliegenthart, R. Studying Dynamic Social Processes with ARIMA Modeling. *Int. J. Soc. Res. Methodol.* **2014**, *17* (6), 693–708.
75. Wang, K. W.; Deng, C.; Li, J. P.; Zhang, Y. Y.; Li, X. Y.; Wu, M. C. Hybrid Methodology for Tuberculosis Incidence Time-Series Forecasting Based on ARIMA and a NAR Neural Network. *Epidemiol. Infect.* **2017**, *145* (6), 1118–1129.
76. Wang, W. C.; Chau, K. W.; Xu, D. M.; Chen, X. Y. Improving Forecasting Accuracy of Annual Runoff Time Series Using ARIMA Based on EEMD Decomposition. *Water Resour. Manage.* **2015**, *29* (8), 2655–2675.
77. Werner, G.; Yang, S.; McConky, K. Time Series Forecasting of Cyber Attack Intensity. In *Proceedings of the 12th Annual Conference on Cyber and Information Security Research*, April 2017; pp 1–3.
78. Willmott, C. J.; Matsuura, K. Advantages of the Mean Absolute Error (MAE) over the Root Mean Square Error (RMSE) in Assessing Average Model Performance. *Climate Res.* **2005**, *30* (1), 79–82.
79. Xiong, Y.; Yeung, D. Y. Mixtures of ARMA Models for Model-Based Time Series Clustering. In *2002 IEEE International Conference on Data Mining, 2002. Proceedings*; IEEE, December 2002; pp 717–720.
80. Yamak, P. T.; Yujian, L.; Gadosey, P. K. A Comparison between ARIMA, LSTM, and GRU for Time Series Forecasting. In *Proceedings of the 2019 2nd International Conference on Algorithms, Computing and Artificial Intelligence*, December 2019; pp 49–55.
81. Zhang, G. P. Time Series Forecasting Using a Hybrid ARIMA and Neural Network Model. *Neurocomputing* **2003**, *50*, 159–175.
82. Zhou, X.; Pan, Z.; Hu, G.; Tang, S.; Zhao, C. Stock Market Prediction on High-Frequency Data Using Generative Adversarial Nets. *Math. Probl. Eng.* **2018**.

CHAPTER 8

Industry Maintenance Optimization Using AI

V. SESHA SRINIVAS[1], R. S. M. LAKSHMI PATIBANDLA[2*],
V. LAKSHMAN NARAYANA[3], and B. TARAKESWARA RAO[4]

[1]*Department of Information Technology, RVR & JC College of Engineering, Andhra Pradesh, India*

[2]*Department of IT, Vignan's Foundation for Science, Technology, and Research, Andhra Pradesh, India*

[3]*Department of IT, Vignan's Nirula Institute of Technology & Science for Women, Andhra Pradesh, India*

[4]*Department of CSE, Kallam Haranadhareddy Institute of Technology, Guntur, Andhra Pradesh, India*

Corresponding author. E-mail: Patibandla.lakshmi@gmail.com

ABSTRACT

Industrial revolution 4.0 has emerged as an ideal scenario for boosting the use of cutting-edge AI and machine learning (ML) services in process control and optimization. The emergence of massive process monitoring data, facilitated by cyber–physical systems distributed alongside manufacturing processes, the proliferation of hybrid Internet of Things architectures powered by polyglot data repositories, including regardless of size data analytics capabilities all are essential components of this latest technological revolution. The Industrial revolution 4.0 methodology shows

Computational Intelligence Applications for Software Engineering Problems. Parma Nand, PhD, Rakesh Nitin, PhD, Arun Prakash Agrawal, PhD & Vishal Jain, PhD (Eds.)

an outsized range of competitive advantages affecting competitiveness, quality, and efficiency primary performance indicators. Since it considered four fundamental measures: availability, quality, overall performance, and overall equipment efficiency have emerged as a target KPI for many manufacturers. This chapter illustrates AI's role in predictive maintenance (PdM), the Need for AI application in Manufacturing, the proposed model for PdM of Industry 4.0, and finally a discussion on accurate results.

8.1   INTRODUCTION

The rise of automation technology in manufacturing is referred to as Industrial revolution 4.0. It is a shift that enables quicker, increasingly versatile, yet reliable using auto data in supply chains. As a result, businesses can manufacture higher-quality products at lower prices. And it is a transition that is not just boosting production and changing the economic landscape. It is accelerating a technical revolution in the process. Machines, however, remain at the heart of the revolt. Although these devices also make our lives easier, their proclivity for breaking down can cause a headache, which can occasionally escalate into a seizure on an industrial scale. Yeah, the vast majority of errors are understandable. Some, though, have a direct effect on human lives. Predictive maintenance (PdM) seeks to prevent such a disaster, but to do so, requires access to large amounts of data. This is now achievable due to increased automation, which is why PdM will transform Industrial revolution 4.0. That protection aspect of a PdM approach comes first after emerges the optimization. It allows production lines to streamline maintenance to prevent future failures, resulting in less overall downtime. As a consequence, it will assist companies in improving computer average performance. The system enables managers to almost fully eliminate the possibility of a breakdown, as well as the negative consequences that such a disruption could bring. As per McKinsey & Company report, PdM will save close to 40% in long-term maintenance costs while also reducing capital expenditures on new machinery and equipment by up to five times. To be accurate, PdM requires a large amount of data, which can be collected using sensors installed in the corporate machinery. These sensors can keep track of several things, including:

Temperature settings: All system temperatures should be monitored, as well as the ambient temperatures of such air.

Humidity (%): Inspecting the heating, ventilation, and air-conditioning systems, and the electrical components.

Intensity: Monitoring parameters such as water pressure in water systems in case of changes or losses that were not expected.

Uniqueness: Since vibration is typically the first sign of a problem, sensors may warn production staff of an imminent failure.

The list goes on, but what matters is not so much what they calculate as it is how to use the data. The majority of businesses use a SCADA framework, which stands for "Supervisory Control and Data Acquisition," as even the name suggests, to collect data, generate visual representations, and help process control. SCADA systems, on the other hand, have several disadvantages. They require human interaction first and foremost: people must configure (and reconfigure) systems for them to operate properly. Furthermore, the systems are unable to detect irregularities in data or predict a sensor reading within a given time frame. Finally, interpreting actionable insights is nearly impossible for a person's operator. As a result, more increasing number of employers are turning to machine learning (ML). Data are analyzed throughout near real-time using ML. Not only is it useful to reach associations in historic experience and real readings, and it is also useful to alert workers about the risk of failure. Even better, no human intervention is required: ML can take care of everything with the aid of statistics.

There will be no more manual tweaking or configuration; instead, let previous output affect subsequent steps because the algorithm:

1. Recognizes possible problems.
2. Prescribes ways to minimize risk.
3. Makes suggestions on the best maintenance schedules.
4. Minimize downtime in all fields.

Expect disruptions while working with robots. At the very least, that is how it will be. After the introduction of automation technologies, it is now possible to virtually eradicate downtime, owing to PdM.[1] PdM detects when something is on the verge of failure and takes action to prevent it. It is fast becoming a pillar of The industrial revolution 4.0 has a wide range of industrial applications. Equipment maintenance is critical in industries, as it affects the kit's running time and performance.

As a result, equipment faults must be detected and corrected to prevent production process shutdown. ML methods have emerged as a promising tool in Application on PdM in prevent machinery failure that makes up the manufacturing floor's assembly lines. Corporations or companies have been reinvented with technology as commercial AI (AI) and therefore the Internet of Things (IoT) increase in size. Companies are learning how to make use of all their data. not only to analyze the past but also to forecast the future. Maintenance may be a key field where significant cost savings and output value are achieved around the world.[2] Globally, $647 billion is lost, according to the International Society of Automation per year due to system downtime. Businesses have overhauled maintenance procedures over the years to reduce downtime such as enhancing quality. However, there continues to be some disagreement about the best approaches for using data when pursuing optimal operational performance. We now can process large quantities with sensors easier than ever owing to AI and computer science. This gives businesses a once-in-a-lifetime opportunity to improve current maintenance activities, while also developing a new idea: PdM.[3–8] Industrial production was the sector that can hope to see unparalleled cost savings as a result of AI. While most manufacturers do use preventive or PdM, AI has the potential to usher in a new age of productivity.

8.2   A ROLE ON ARTIFICIAL INTELLIGENCE IN TOTAL PRODUCTIVE MAINTENANCE (TPM)

TPM is a comprehensive approach to maintenance toward maintaining and improving essential assets and operating practices that result in fewer disasters, reduced downtime, increased efficiency, and improved safety.[9–15] Some industrial companies just use this method, which was created in the 1960s, to proactively execute equipment maintenance based on historical data and resource constraints for when repairs are predicted. TPM optimizes overall equipment effectiveness (OEE) and plant efficiency by employing planned maintenance concepts. Maintenance cost will help you avoid breakdowns and increase the useful life of your assets. Maintenance with Artificial Intelligence and Autonomy TPM's Autonomous Maintenance is a key function that requires adoption and implementation (AM). Everyone is responsible only for machine productivity but service in this type of maintenance. Instead of relying solely on maintenance personnel to restore assets, machine operators are in charge of their maintenance. Experts were released focused on broader modifications that improve system efficiency by making machine operators perform routine maintenance on assets. Because it needs a significant quantity of awareness and consultation, AM is notoriously difficult to implement.

Machine operators lack historical machine expertise that technicians possess, and technicians will be unable to accept certain activities when they have willpower into future job responsibilities. Businesses will now benefit from AI-driven applications that make AM attainment at ease. Battlefield operators now have a greater understanding of their computers than always earlier. Having all previous data in one easy-to-access dashboard keeps the entire firm on the same page and speeds up computer service. Businesses will now ensure that each operator has the appropriate equipment and, as a result, the required skills at the essential stage to complete the job.

PdM vs. Planned Preventive Maintenance:

PPM, or scheduled preventive maintenance, is maintenance that is initiated by period or incidents that necessitate repair. This type of system, which would be a critical element of TPM, ensures that maintenance is scheduled when machines are still operational, avoiding unplanned downtime, and maximizing the equipment's lifetime and efficiency. The current system, though efficient, has some disadvantages. It is not a precise science, and it can lead to over-or under-maintenance of possessions. It also relies on guidance for routine inspections that are devoid of experience.

PdM uses condition-based indicators and alarms to optimize the upkeep cadence and improve vehicle accessibility even as the vehicles are around to disruption. Unless the vehicle is at risk of overloading outside of the scheduled maintenance schedule, such example, the car will alert the user. Such servicing becomes conducted preemptively while vehicles are still running but are at high risk of failure.

Most companies are moving to predictive and condition-based maintenance, which is powered on artificial intelligence and statistics, as the sector has become more integrated. Knowledge management is becoming simpler but more frequent even as sector grows more networked. PPM is largely concerned with information that is time-based. When it comes to automobiles, for example, the amount of time spent driving or the number of miles driven determines maintenance.

These data frequently compares the performance of an existing property to the performance of the entire portfolio. Data essentially forecasts what will happen. The bulk of maintenance systems concentrate on data transfer rather than congestion control for real-time analytics. However, sending the information is the first step; what does with it after that is what matters. AI and ML will assist and in quicker aggregation and use of data.[16–18]

PdM uses data from several sources, including historical maintenance reports, computer sensor data, afterwards, weather data are being used to govern when a gadget will need to be serviced. After combining huge benefit statistics and time series, clients would be able to make informed choices on whether it is necessary to maintain the system. PdM aspires with precision agriculture software to convert vast volumes of data into specific insights and data sets, preventing statistics overload.

Huge amounts with uncertain data can now be easily extracted for further importance with sensor ML algorithms models. PdM tools that use artificial intelligence can enhance current maintenance processes, ensuring that people have the expertise and tools they need to keep mission-critical assets working at their best. Its advantages to the customer are as follows: Maintenance costs are reduced by around 50%, accident failures are reduced by around 55%, overhaul and repair time is reduced by 60%, additional storage is reduced to around 30%, and the mean solar time between failures is boosted by 30%.

Although these can seem to be miraculous changes at first glance, keep in mind that these are alleged research numbers; in fact, the chances are lower since each case is unique. Despite the fact also that chances have been the evidence of improvement was also minimized for the company would be apparent. Also, if prior maintenance costs were huge, a massive amount would be avoided. For example, a typical manufacturer can save up to 10% on maintenance costs, which is the same of a 40% emphasis on quality. The four most widely used ranges on signs of a cyber-attack could be identified by business potential failure or mistake that will affect their company or possibly any small processes in it using predictive and analyzing large datasets in maintenance.[19–23]

While addressing manufacturing and the IoT, some words that come to mind are manufacturing and the IoT. GEMÜ, the top automation component and valve manufacturers, designed a technology-enabled IoT to monitor the capability of production processes to execute even to an extend that malfunctioning parts may be identified and discarded before they fail, optimizing cost effectiveness.

Also in the automotive industry, we now have connected cars that generate a vast amount of data from sensors mounted close to the vehicle. The information is then sent to car suppliers and dealerships, who alert drivers to any problems and help them get to a mechanic prior to the car explores.

Utility companies use PdM to better manage their internal operations by monitoring massive amounts of data generated by smart meter in order to detect early symptoms of supply and demand problems and take proactive measures to avoid outages. Insurance companies could earn via PdM while also enhancing their severe weather predictive analytics.

8.3   NEED FOR AI APPLICATION IN MANUFACTURING

When it involves applying ML to manufacturing use cases to enhance OEE, there are some ways to deal with the business problem and lots of algorithms to use. This section analyzes the OEE metric and appears at how ML is currently employed by manufacturers to enhance availability, performance, and quality.

Improved availability of Recall that the available OEE estimation is:

Availability = (Planned Prod. Time – Stoppage Time)/Planned Production Time

Decrease Stoppage Time and be more precise in forecasting Planned Production Time to boost the presented score. Predictions to minimize stoppage time and forecasting to improve production time planning are both covered in this section.[24] Stoppage time is often planned (machine setup time) and unplanned (either a machine or asset failure, or maybe even a rejection during the manufacturing operation with the part that requires resolution or a reset) in the OEE calculation. We focus on unscheduled halt duration in either part because it is frequently the most critical variable and where ML can be used to make predictions. PdM, for example, can be used to prevent unplanned machine downtime; however, a similar concept is frequently used to predict faults in manufacturing processes. PdM reduces unplanned downtime by allowing enables proactive planning and execution of machine or asset repairs reoccurrence of a failure.[25]

Two approaches are regularly used to achieve forecasting, both of which rely on failure events proving few. And provide further context, a CNC machine's final sample size to such a major collapse of a component within the hydraulic ram could be difficult to foresee if historical sensor data cannot indicate leading indicators that it would fail or has failed in the past. And from the other hand, if previous current spikes have shown that a motor could be about to fail, future failures are often predicted. This recent increase, in this case, is caused to a proxy event. Other events could be used as a proxy for failures if the failure event itself is not or cannot be captured to some extent. Maintenance logs, repair bills, and machine operator run logs, for instance, might be used if a proxy event is close enough to the major failure event to allow the prediction to be validated. If failure is not prevalent, supervised learning is regularly used. In these situations, a model is trained using historical records of maintenance events using time-series telemetry records of leading indicators such as vibration, temperature, and pressure. Tabulated categorical classifiers, such as XGBoost, are the major model family we recommend trying.

The major drawback of such models is that feature engineering will be required, and this will vary depending on the device.[26] Relationships between indicators, for example, must be manually checked or contributed to a dataset by SMEs, such as systems administrators. However, results can often be obtained with less data, model training is faster, and the model is thus (relatively) easier to understand. Deep learning architectures with sequential input, like recurrent neural networks, would form the second model family (RNNs). The main disadvantages of those models are that the larger dataset requires additional training periods, making the model easier to grasp. But with deep learning, some level of feature engineering is frequently expected. When failure events are few, unsupervised learning is commonly used. Within those situations, real-time data augmentation is commonly done on the index itself, providing engineers with early notice of possible device issues. For example, if a replacement sensor is installed, past data for that sensor is no longer available. If domain expertise demonstrates that heavy vibration is associated to device failures, outlier occurrences in the new vibration sensor should be sent to the engineer, potentially identifying a failure before it develops. The Random Cut Forest technique is one of the most common outlier detection methods used on time-series datasets.

Improved production time planning with demand forecasting allows for more precise estimates of planned production time by anticipating the quantity of a product that needs to be produced. Forecast models are created from historical time-series datasets of product demand to do demand forecasting. There are numerous time-series forecasting solutions available, and trade-offs are always considered. Traditional approaches, such as that of the autoregressive integrated moving average, require more input because they can only model one statistic at a time (e.g., one SKU at one customer). However, because the model ignores other associated statistic datasets, like vacations, forecasts may be reduced. Cutting-edge deep learning-based models, such as DeepAR, draw connections across different time-series (e.g., between product lines and consumers) should provide high forecast accuracy, but this requires the algorithm to get a lot of knowledge. Additional business criteria, such as forecast horizon, product resolution (SKU vs product line), customer resolution (vendor vs channel vs area), and so on, get calculated to establish a ratio between forecast granularity and performance.

Improve performance as mentioned within the introduction, the OEE calculation on behalf of performance is

$$Performance = (Cycle\ Time \times Cycles)\ Run\ Time$$

As a result, one must dramatically reduce Bytecode and otherwise boost Cycle to enhance the performance index. Unless the approach is adjusted or updated, Cycle Time will not change. Process optimization boosts productivity by improving method variables to increase production speed while keeping errors under control. The goal of completely auto-mated process optimization is accomplished in stages. From a practical sense, any device provides benefits and weaknesses that must be examined and balanced. The ideal aim would be a machine that functions without any human interaction, but we will go over each step on the way to getting there below.

Before we get into any discussions about applying ML for process optimization, it is important to note that simulates can all be immediately replaced by ML. Because the calculations they create are backed by physical and chemical properties, parameters calculated from the first principles should be employed for the initial optimization. ML, on either hand, may be more useful in accounting for imperfections that distinguish physically simulation leads than real-world operational settings. The two should be used together.

After that, I remarked, "Let's start talking about process optimization." We would like to establish guardrails for the approach before improving its settings. For example, before recommending that what a machine's route pace be increased, a modeling first must verify that the proposed line speed will not result in defective parts or unscheduled stoppages due to low quality. This quality prediction application is addressed in greater detail in the Quality Metric section later. Simply said, a high-quality prediction model should always be created then use historical failures with supporting data such as set points and environmental sensor readings.

With this quality prediction model in place, the user can start actively refining some procedures.[27] For example, an operator could first enter a setting that they believe will boost operating speeds while still satisfying the required specifications. Whenever this proposed setting is paired with other settings and sensor readings, the model can assess if it will fail. If there is a significant risk of failure, the operator is warned and given the option to manually try again with a lower value. Following that, regression models are frequently trained to forecast the method's cycle time.

The data were the same as before, however, the focus will be on variety (e.g., components per, throughput, or flow) rather than a single category (e.g., pass/fail). This technique optimization was inspired by this regression model. The operator can validate new settings to increase the method or machine speed while maintaining within the technique edge protection whenever the two models were currently running. For example, the operator can predict an increase during a set, and the ML models may predict that this setting will result in a faster line speed and higher-quality parts.

The operator can then put this setting to the test. However, because the models are trained via supervised learning, they will only actualize particular possibilities if they have seen the scenario before in historical data. Historic fault info will not exist or may be restricted in number in only certain edge cases and where research is expensive. Similarly, previous data on machine performance and, as a result, the associated settings may be conservative, and so data for a certain maximum speed may not exist.

Because the model has never experienced these situations before, it will be unable to predict some failures or prescribe the best speed. To be able to use ML models for automated process control, a knowledge set must be constructed that includes virtually all scenarios so that the model can identify and make the correct with a high level of certainty, it is impossible to make a prediction.

Until then, SMEs classically use the models that can safely fine-tune and experiment with settings, combining the basics of both human expertise and ML forecasts. The regression model will learn how each setting affects cycle time over time.[28] The SMEs can manually experiment with the most mathematically significant settings while confirming that the results are consistent with their prior experiences. Method optimization algorithms, such as Bayesian Optimization or the reinforcement learning model, will experience more scenarios with time, increasing the accuracy with its prescriptions of optimum settings. When the model has observed enough cases, it is frequently used for a process optimization solution that may dynamically transfer setting to the machinery simply a set of inputs, resulting to maximum machine or line performance.

Improve quality as with the opposite OEE variables, quality is calculated with:

$$Quality = Good\ Count/Total\ Count$$

Before we get into the mechanics of quality predictions, it is important to note that improving quality is not the same as detecting flaws after the fact. The goal of enhancing quality is to ensure that the production processes' variable controls are set such that low-quality products are not made in the first place.

By transferring one or more inspection jobs to automated systems, automated quality testing reduces the time spent performing slower manual inspections. Automated quality testing is a type of human augmentation in manufacturing, in which routine or basic operations that were previously performed by teams of highly trained humans are sometimes delegated to an algorithm, allowing these experts to focus on important tasks. To carry out certain enhanced quality testing, the pass/fail decisions made by humans, as well as the supporting indicators that standard inspectors want to employ in giving their opinion, must now be documented. The dataset is then used to create a classification model that will classify future parts/products as pass or fail automatically. These categorization models can be delivered in a variety of formats, depending on how the dataset is structured.

This section discusses two types of formats that are frequently used in quality testing: tabular and pictorial. Rectangular dataset is the most prevalent data format for quality analysis since it contains the most relevant data. Electronic database tables containing prior pass/fail records are connected with relevant supporting indicator tables and exported as if they are available. models with csv files in a knowledge lake When good databases are not accessible, spreadsheets containing manually-tracked entries are frequently exported to just the data lake (although it was not usually advised since it is prone to human mistake and is not scalable).

The.csv files that result should include at least three columns: time-stamp, device ID, and pass/failure record. Other columns should offer supporting signs that the standard inspectors will use to determine whether something is pass or fail. These indicators vary by product, however, they could include weight, size, hardness, as well as other features. Following the data acquisition, feature engineering is typically employed to enhance the statistics to match the inspector's manual method, such as translating weight and dimensions to density. These historical records can then be sent into a tabular classification model like XGBoost to be trained. Although image data are not as common as tabular data, it has recently received a lot of attention as a result of recent computer vision research and applications. Previous human-labeled pass/fail data are merged with images taken during the inspection by conventional inspectors in certain application scenarios. These labels are commonly found only at neighbor or fault level (e.g., the entire circuit board) (e.g., scratches and dents). Feature engineering is most commonly used on raw images to highlight the features that inspectors use to make judgments.

These procedures could encompass eliminating areas of the image that are not needed, refining edges, boosting color, and more. For help its shall be granted better, image augmentations such as flipping, rotation, and brightness/contrast changes are frequently used. Successful completion of these preprocessing stages, a model is trained on the texture features. Deep learning based image classification architectures with pretrained models, such as the VGG or Resnet families, are exploited to part-level classifications. Object detection architectures, including pretrained models like SSD (Single Shot Detection) or Yolo, can be used for defective level classifications.

However, keep in mind that pretrained models are typically trained with commonplace items like kittens and dogs, which may not translate to your specific photographs of manufacturing parts or assemblies. The use of computer vision for quality control is not limited to pass/fail whether it comes to the standard of an area or a product. Computer vision could also be used to help with product sort and quality classification. By assigning two or many classifications into automation systems, this exercise can reduce execution time.

This use case is similar to augmented quality testing, but instead of a simple pass/fail decision, one or more classes, such as grades, sizes, or downstream treatments, are frequently assigned. A technology similar to those used in augmented quality testing is frequently used here. The only difference is that the labels are now multiclass rather than binary, and they will be used to sort the standard of raw items, such as wood or food, into the numerous categories that are used in the manufacturing processes for the top product.

Through analyzing operations KPIs, quality processes prediction raises the great item count by detecting and remediating possibly problematic items from the primary development process. For example, monitoring the pressure and temperature of a production line machine cast and informing the user if faulty parts will be produced until the mold's recipe is corrected. The application of quality predictions is quite similar to PdM if the forecast has been on the part rather than the machine, and thus the manufacturing process is remediated rather than the machine being maintained. Otherwise, the methods are similar to those used in PdM. It is preferable to have a labeled time-series dataset of jump inspections across many phases of production, as well as leading indications to support this pass/fail inspection. The presence of this labeled dataset necessitates the use of ML techniques such as XGBoost or RNNs for supervised learning. When a labeled time-series dataset was unavailable, for as when upstream problems are uncommon, an unsupervised technique is frequently used. Leading indicators are tracked over time, and anomalies in these indications are detected and reported to engineers, comparable to PdM. Random Cut Forest may have been an algorithm in this heuristic search instance.

8.4   PROPOSED MODEL

Demand forecasting is used in several sectors, including retail, e-commerce, manufacturing, and transportation. It uses historical data to feed ML algorithms and models to forecast the number of goods, facilities, power, and other factors. It enables companies to capture and process data from across the entire supply chain, decreasing costs and increasing productivity. Demand forecasting powered by ML is extremely accurate, fast, and transparent. Businesses may extract actionable insights from a steady stream of supply/demand data and respond to changes as required (Fig. 8.1).

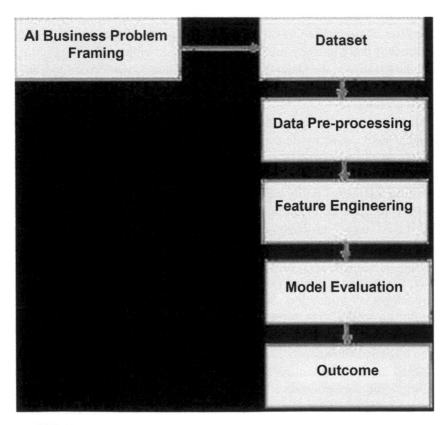

FIGURE 8.1   Model process flow.

Figure 8.1 represents the process flow of the proposed model, in that first identify the business problem and apply it to any dataset. After that preprocessing the data then feature selection and extraction. Next to creating and evaluating the model and then deploying it (Fig. 8.2).

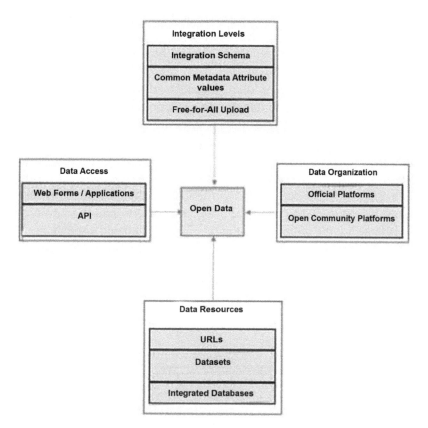

FIGURE 8.2   Unified AI model implementation.

Figure 8.2 represents the Unified AI Model Implementation process. Here, Open Data are the platform and communicates with the different levels of data like collecting the data from various resources, organization of data by using official, and open community platforms, Integration levels such as schema, common metadata attribute values, and access the data from web forms and applications.

8.5   RESULTS AND DISCUSSIONS

This analysis had an AI4I 2020 PdM dataset, which could be a virtual dataset that shows practical PdM data collected mostly in the industry using Recurrent Neural Networks of Tensor Flow. Multiple multivariate statistics with "cycle" as the unit of time, as well as 25 products air temperature readings for each cycle, make up the training results. Whenever series are assumed to be created from the air temperature about a comparable category of the product. Because of the training data, any testing data have an associated data model. The only distinction is that the information does not define after the disaster arises. As a final point, the powdered truth data indicate how many operating cycles the research data have left. Regression Model:

The following figures show that the loss function's tendency, the Mean Absolute Error, R^2, and actual data compared predicted data. This table represents the Mean Absolute Error rate and determination coefficient values when using the regression model (Figs. 8.3–8.6 and Table 8.1).

TABLE 8.1   Regression Model.

Mean absolute error	12
Determination coefficient (R^2)	0.7965

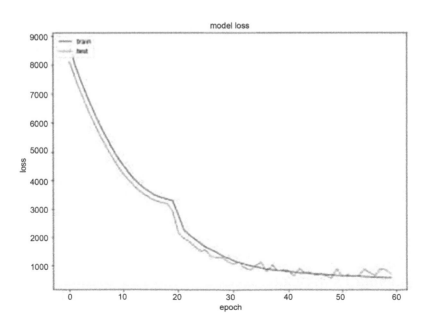

FIGURE 8.3   Regression model loss.

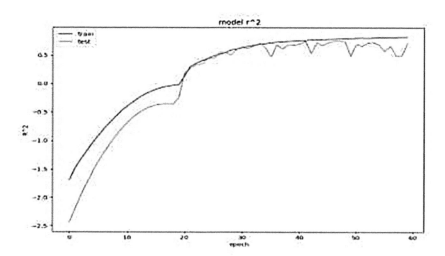

FIGURE 8.4   Regression model coefficient.

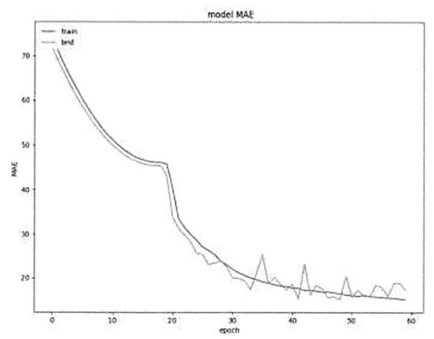

FIGURE 8.5   Regression model MAE.

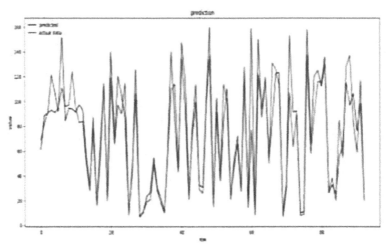

FIGURE 8.6   Regression model prediction.

Binary Classification: The following figures show the desire to lose actual data that are compared to predicted data in terms of function, accuracy, and accuracy.

The following table shows the binary classification representation (Figs. 8.7–8.9 and Table 8.2).

TABLE 8.2   Binary Classification.

Accuracy	0.97
Precision	0.92
Recall	1.0
F-score	0.96

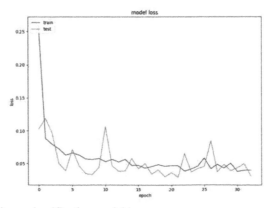

FIGURE 8.7   Binary classification model loss.

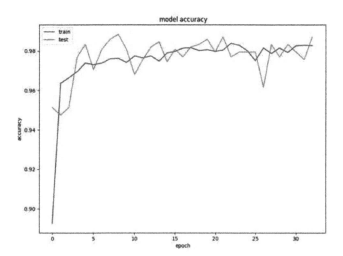

FIGURE 8.8   Binary classification model accuracy.

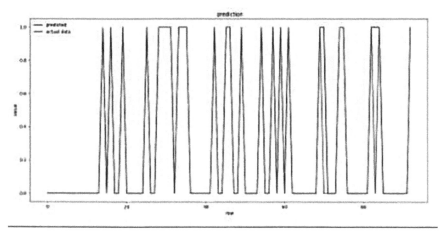

FIGURE 8.9   Binary classification model prediction.

8.6   CONCLUSION

PdM has been used in only manufacturing and automotive industries, but it is most widely used in those two. This software not only lowers maintenance costs, but also reduces unforeseen losses, redesign, or corrective maintenance across nearly 60%, resulting in dramatically higher equipment and computer uptime. Manufacturing executives are beginning to notice the value of combining PdM with ML to operate expensive and complicated machines, and Industrial Revolution 4.0 would agree. While the adoption of PdM remains novel, only top AI companies can handle the appropriate application of the technology. This might be one of the dollar's leaders when it comes to such usage of such a technology!!

KEYWORDS

- **machine learning**
- **artificial intelligence**
- **smart manufacturing**
- **classification**
- **cyber-physical systems**
- **regression**
- **data-driven**
- **industry 4.0**
- **predictive analytics**

REFERENCES

1. Hedman, R.; Subramaniyan, M.; Almström, P. Analysis of Critical Factors for Automatic Measurement of OEE. *Procedia CIRP* **2016,** *57*, 128–133. DOI: 10.1016/j.procir.2016.11.023

2. Ting, K. M. Confusion Matrix. In *Encyclopaedia of Machine Learning and Data Mining*; Springer: Boston, USA, 2017. DOI: 10.1007/978-1-4899-7687- 1_50

3. Huggett, D. J.; Liao, T. W.; Wahab, M. A. et al. Prediction of Friction Stir Weld Quality without and with Signal Features. *Int. J. Adv. Manuf. Technol.* **2018,** *95* (5–8), 1989–2003. DOI: 10.1007/s00170- 017-1403-x

4. Zheng, A.; Casari, A. *Feature Engineering for Machine Learning: Principles and Techniques for Data Scientists*; O'Reilly Media Inc: Sebastopol, USA, 2018. ISBN: 1491953195, 9781491953198

5. Bonada, F.; Echeverria, L.; Domingo, A.; Anzaldi Varas, G. AI for Improving the Overall Equipment Efficiency in Manufacturing Industry, 2020. 10.5772/intechopen.89967.

6. Li, Z.; Wang, K.; He, Y. Industry 4.0—Potentials for Predictive Maintenance, 2016. 10.2991/iwama-16.2016.8.

7. Lee, W. J.; Wu, H.; Yun, H.; Kim, H.; Jun, M. B. J.; Sutherland, J. W. Predictive Maintenance of Machine Tool Systems Using Artificial Intelligence Techniques Applied to Machine Condition Data. *Procedia CIRP* **2019**, *80*, 506–511. ISSN 2212-8271.

8. Yan, J.; Meng, Y.; Lu, L.; Li, L. Industrial Big Data in an Industry 4.0 Environment: Challenges, Schemes, and Applications for Predictive Maintenance. *IEEE Access* **2017**, *5*, 23484–23491. DOI: 10.1109/ACCESS.2017.2765544.

9. Nguyen, K. A.; Do, P.; Grall, A. Joint Predictive Maintenance and Inventory Strategy for Multi-Component Systems Using Birnbaum's Structural Importance. *Reliab. Eng. Syst. Saf.* May **2017**, *168*, 249–261. DOI: 10.1016/j. ress.2017.05.034.

10. Ladj, A.; Varnier, C.; Tayeb, F. B. S. IPro-GA: An Integrated Prognostic Based GA for Scheduling Jobs and Predictive Maintenance in a Single Multifunctional Machine. *IFACPapersOnLine* **2016**, *49*, 1821–1826. DOI: 10.1016/j.ifacol. 2016.07.847.

11. Okoh, C.; Roy, R.; Mehnen, J. Predictive Maintenance Modelling for Through-Life Engineering Services. In *The 5th International Conference on Through-life Engineering Services*, Vol. 59; The Author(s), 2017; pp 196–201. DOI: 10.1016/j. procir.2016.09.033.

12. Schmidt, B.; Wang, L. Cloud-enhanced Predictive Maintenance. *Int. J. Adv. Manuf. Technol.* **2018**, *99* (1–4), 5–13. DOI: 10.1007/ s001700168983-8.

13. Lindstrom, J.; Larsson, H.; Jonsson, M.; Lejon, E. Towards Intelligent and Sustainable Production: Combining and Integrating Online Predictive Maintenance and Continuous Quality Control. In *The 50th CIRP Conference on Manufacturing Systems*, Vol. 63; The Author(s), 2017; pp 443–448. DOI: 10.1016/j. procir.2017.03.099.

14. Schmidt, B.; Wang, L.; Galar, D. Semantic Framework for Predictive Maintenance in a Cloud Environment. In *10th CIRP Conference on Intelligent Computation in Manufacturing EngineeringCIRP ICME '16*, 2017; Vol. 62, pp 583–588. DOI: 10.1016/j.procir.2016.06.047.

15. Chiu, Y. C.; Cheng, F. T.; Huang, H. C. Developing a Factory-Wide Intelligent Predictive Maintenance System Based on Industry 4.0. *J. Chinese Inst. Eng.* **2017**, *40* (7), 562–571. DOI: 10.1080/02533839.2017.1362357.

16. Wang, K. Intelligent Predictive Maintenance (IPdM) System, Industry 4.0 Scenario. *WIT Trans. Eng. Sci.* **2016**, *113*, 259–268. DOI: 10.2495/IWAMA150301.

17. Yamato, Y.; Fukumoto, Y.; Kumazaki, H. Predictive Maintenance Platform with Sound Stream Analysis in Edges. *J. Info. Process.* **2017**, *25*, 317–320. DOI: 10.2197/ ipsjjip.25.317.

18. Traore, M.; Chammas, A.; Duviella, E. Supervision and Prognosis Architecture Based on Dynamical Classification Method for the Predictive Maintenance of Dynamical Evolving Systems. *Reliab. Eng. Syst. Saf.* **2015**, *136*, 120–131. DOI: 10.1016/j. ress.2014.12.005.

19. Baidya, R.; Ghosh, S. K. Model for a Predictive Maintenance System Effectiveness Using the Analytical Hierarchy Process as Analytical Tool. *IFAC-PapersOnLine* **2015**, *28*, 1463–1468. DOI: 10.1016/j.ifacol.2015.06.293.

20. Lei, X.; Sandborn, P. A.; Goudarzi, N.; Bruck, M. A. PHM Based Predictive Maintenance Option Model for Offshore Wind Farm O&M Optimization. In *Proceedings of the Annual Conference of the Prognostics and Health Management Society 2015*, 2015; pp 1–10.

21. Krupitzer, C.; Wagenhals, T.; Züfle, M.; Lesch, V.; Schäfer, D.; Mozaffarin, A.; Edinger, J.; Becker, C.; Kounev, S. *A Survey on Predictive Maintenance for Industry 4.0*, 2020.

22. Bousdekis, A.; Lepenioti, K.; Apostolou, D.; Mentzas G. Decision Making in Predictive Maintenance: Literature Review and Research Agenda for Industry 4.0. *IFAC-PapersOnLine* **2019**, *52* (13), 607–612. ISSN 2405-8963.

23. Wuest, T.; Weimer, D.; Irgens, C.; Thoben, K. D. Machine Learning in Manufacturing: Advantages, Challenges, and Applications. *Prod. Manuf. Res.* **2016**, *4*, 23–45.

24. Hakeem, A. A. A.; Solyali, D.; Asmael, M.; Zeeshan, Q. Smart Manufacturing for Industry 4.0 Using Radio Frequency Identification (RFID) Technology. *J. Kejuruter* **2020**, *32*, 31–38.

25. Jiang, Y.; Yin, S. Recursive Total Principle Component Regression Based Fault Detection and Its Application to Vehicular Cyber-Physical Systems. *IEEE Trans. Ind. Info.* **2018**, *14*, 1415–1423.

26. Shin, J. H.; Jun, H. B.; Kim, J. G. Dynamic Control of Intelligent Parking Guidance Using Neural Network Predictive Control. *Comput. Ind. Eng.* **2018**, *120*, 15–30.

27. Falamarzi, A.; Moridpour, S.; Nazem, M.; Cheraghi, S. Prediction of Tram Track Gauge Deviation Using Artificial Neural Network and Support Vector Regression. *Aust. J. Civ. Eng.* **2019**, *17*, 63–71.

28. Scalabrini Sampaio, G.; de Vallim Filho, A. R. A.; da Santos Silva, L.; da Augusto Silva, L. Prediction of Motor Failure Time Using an Artificial Neural Network. *Sensors* **2019**, *19*, 4342.

CHAPTER 9

Comparative Study of Invasive Weed Optimization Algorithms

SHWETA SHRIVASTAVA*, D. K. MISHRA, and VIKAS SHINDE

Department of Engineering Mathematics & Computing,
Madhav Institute of Technology & Science, Gwalior, India

Corresponding author. E-mail: Shwetashrivastava986@gmail.com

ABSTRACT

In this chapter, the efficacy of Invasive Weed Optimization (IWO) is evaluated via a collection of multimodel benchmark functions, such as global and local minima which are identified by IWO. IWO algorithm is a population-based algorithm approach stimulated by the behavior of weeds colonies. Unwanted plants (weeds) are those plants that are energetic, and their intrusive expansion mannerism is a crucial hazard to farming. Weeds have proven highly resilient and adaptable to environment change. Capturing their properties will therefore result in a powerful optimization algorithm. Easy but efficient optimizing algorithm known as IWO is attempted to imitate the hardness, acclimatization, and haphazardness of colonizing weeds. IWO was tested on 10 benchmark functions. Lastly, statistical comparisons of results were made with other methods, showing that IWO performs better than other algorithms.

Computational Intelligence Applications for Software Engineering Problems. Parma Nand, PhD, Rakesh Nitin, PhD, Arun Prakash Agrawal, PhD & Vishal Jain, PhD (Eds.)
© 2023 Apple Academic Press, Inc. Co-published with CRC Press (Taylor & Francis)

9.1 INTRODUCTION

A weed is an unwanted plant that may be endemic to the area but detrimental to the crop being farmed.

Farmers have adopted numerous methods to control weeds, but all of them have their shortcomings. Physical and mechanical methods are often labor-intensive and are not fiscally prudent. Maclaren et al.[1] suggested the methods like manual removal, tilling, thermal methods, crop rotation, animal grazing, etc. Robertson and Grace[2] informed the increasing global warming and pollutions, physical methods such as the burning of weeds are discouraged. Other methods of weed control involve using chemical, which decreases the fertility of the soil, which adds to the cost and may render the land barren. Moorman[3] proposed the harsh chemicals like glyphosate trifluralin, clethodim, etc., negatively affect the soil microfauna, further contributing to the barrenness of the land. These harmful chemicals ultimately get involved in the human food chain through continuous development accumulation via the organisms of lower trophic levels. Hoy et al.[4] obtained these chemicals then get into human consumption and cause slow, steady effects to develop chronic disease like respiratory damage, cancer, congenital malformations, autism to name a few. The use of weedicides and herbicides is also detrimental to the bird population since birds feed on insect loaded with these chemicals. Therefore, a novel method that can help elucidate and predict the spread of invasive weed is warranted. Invasive Weed Optimization (IWO) is a biological stimulated, metaheuristic optimization method, which emulates the natural behavior of unwanted plants during colonizing. Meta-heuristic algorithms are iterative and nature-based algorithms. Metaheuristic word is made by two individual words "meta" and "heuristic" meaning "higher dimension" and "to obtain or to explore," respectively. These are used to construct the formation in sequence to obtain effectively optimal solution. There are so many meta-heuristic algorithms such as GA, DE, PSO, BBO, FPO, MVO, IWO, and others. A successive iteration process which demonstrates a several view for exploiting and exploring the search space, learning strategies are applied to construct the information in sequence to obtain effectively approximate optimal results. Mehrabian and Lucas proposed IWO,[6] used to solve technical problems and analyzed the outcomes to other algorithms. The results obtained from IWO are better as compared with other algorithms. IWO has also been successfully used to solve various engineering problems. Mainly three algorithms have been used for comparison with Artificial Bee Colony (ABC) (Karaboga and Basturk, 2007), Genetic Algorithm (GA) (Holland 1992), and Particle Swarm Optimization (PSO) (Eberhart and Kennedy, 1995). Researchers have analyzed the performance of IWO. Some studies have been carried out to solve problems with mathematical optimization problems.[9] Zhang et al.[7] used the concept of the heuristic algorithm to build the IWO. They implemented the IWO by the method of crossover and approved the neoteric manner on statistical level, comparing the solutions of the evolved IWO with the results of the levelled IWO and PSO. Chen et al.[8] developed the relation between IWO algorithms for solving permutation flow-shop scheduling problem. Barisal and Prusty[11] applied the IWO to solve large-scale economic problems with a goal of reducing the costs of manufacturing and transferring products subject to limitations on production, consumer request and marred due to the products intervening teleportation and mitigating another disaster. Liu and Wu[10] proposed synthesis of thinned array by IWO. Zhao et al.[12] depicted a vehicle routing problem by discrete hybrid IWO. Yaseen et al.[13] obtained the solution of Global Optimization Problems with hybrid IWO Algorithm and Chicken Swarm Optimization Algorithm. Rad and Lucas[14] proposed a recommendation of system based on IWO algorithm. Mallahzadeh et al.[15,16] applied the IWO technique for antenna configurations. For instance, Zhang et al.[17] introduced a novel optimization-based access to use IWO algorithm encoding sequences for DNA computing. Jatoth et al.[22] proposed optimal suitability of cloud service composition using modified IWO method that is gratifying the balancing of quality of service criterion as well as connectivity conditions of cloud service composition. Uyar and Ulker[23] informed that the IWO is an agriculture-based phenomenon which inspired from ecological happening for evaluation of B-spline curve fitting. Rahmani et al.[24] informed the method which has been stimulated by pliable, defiant, and intrusive comportment of unwanted plants in examine in order to resources and suitable locale for expansion and replica in the surrounding Basak et al.[25] provided a very effective mongrel evolutionary method such implants the dissimilar transmitter which is based metamorphosis of Differential Evolution (DE) into IWO. Falco et al.[26] access is applied Differential Evolution as the local algorithm and determined on a large set of traditional benchmark functions against a large quantity of succeeding. Ghalenoei et al.[27] developed discrete invasive weed optimization (DIWO) algorithm and comparative study with other algorithms. Giri et al.[28] provided the solution of modified IWO accomplishes as good as other competitive actual parameter of optimization and can be applied in justification of optimal value. Kundu et al.[30] presented the fundamental IWO for taking on the multiobjective optimization problems that acquiring more than one objective with concept of fuzzy dominance has been applied to categorize the promising candidate solutions. Li et al.[31] studied to optimization of the Yagi-Uda antenna for the uppermost direction by IWO.

This chapter represents an uncomplicated numerical optimization algorithm is tried to be implemented, stimulated by unwanted plants colonization known as IWO. The IWO method is an uncomplicated but it has proved effectual in converging to the best possible solution by using growth, seeding, and rivalry in a colony of weeds. Consistency and efficiency of suggested algorithm are determined by simulation studies.

9.2   INVASIVE WEED OPTIMIZATION

The method is explained in the following manner:

Step 1 (Initialize a population): A population is initially generated by the distribution of seeds with random locations atop the exploration field.

Step 2 (Reproduction): Each weed of population can generate some seeds, depending on their appropriateness efficacy. Greater the appropriateness of the unwanted plants (weeds), more seeds it produces. The weeds producing formula is

$$weed_i = \frac{f_{max} - f_i}{f_{max} - f_{min}} \times (S_{max} - S_{min}) + S_{min}$$

Where f_{max} = best fitness value, f_{min} = worst fitness values of weed of the present population. S_{max}, S_{min} = maximum and minimum number of seeds to be produced for each weed, respectively. f_i = fitness efficacy of ith weed.

Step 3 (Spatial dispersal): The produced seed spread randomly atop the exploration field with a normal distribution that has mean zero and varying standard deviation σ. However, σ reduces from s_{max} to s_{min} in every generation. The standard deviation for the kth iteration σ_k is given by

$$\sigma_k = \left(\frac{iter_{max} - k}{iter_{max}} \right)^n \times \left(\sigma_{max} - \sigma_{min} \right) + \sigma_{min}$$

where k = current iteration number, n = nonlinear modulation index, $iter_{max}$ = maximum number of iteration.

Step 4 (competitive exclusion): The seeds have obtained their location atop search location; the weeds will be developed by new seeds (Fig. 9.1).

9.2.1   INVASIVE WEED OPTIMIZATION ALGORITHM IN GENERAL

Instate M operators each with m memory or M factors for one solution

Begin i=1 to Iteration count For Loop

Begin k=1 to Iteration End

Begin J=1 to N for Loop [N is the quantity of mother weed agents]

Select solution subset for agent j where solution subset $\in \{1, 2... M\}$

Ascertain the spatial parameters in view of wellness parameters

Perform Spatial Dispersal

Perform Reproduction to produce new weed from mother weed

Perform competitive exclusion to compensate for the excess weed operators

End j for Loop

Assess Fitness for every Weed

Upgrade Global Best if Better Weed Found

Supplant the less fitted ones with the more fitted ones.

End k for Loop

End i for Loop

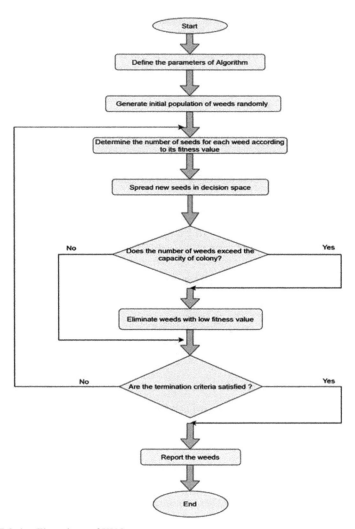

FIGURE 9.1   Flowchart of IWO.

TABLE 9.1   The Benchmark Test Functions.

Sr. no.	Function name	Test function	Range	F (min)
1	Easom	$f(t) = -\cos(t_1)\cos(t_2)\exp[-(t_1 - \pi)^2 - (t_2 - \pi)^2]$	$(-100,100)$	0
2	Matyas	$f(t) = 0.26(t_1^2 + t_2^2) - 0.48\,t_1 t_2$	$(-10,10)$	0
3	Rastrigin	$f(t) = \sum_{i=1}^{m}\left[t_i^2 - 10\cos(2\pi t_i) + 10\right]$	$(-5.12,5.12)$	0

TABLE 9.1 *(Continued)*

Sr. no.	Function name	Test function	Range	F (min)
4	Goldstein-Price	$= [1 + (t_1 + t_2 + 1)^2](19\text{-}14\,t_1 + 3\,t_1{}^2 - 14t_2 + 6t_1 t_2 + 3\,t_2{}^2)]*[30 + (2t_1 - 3t_2)^2\,(18\text{-}32t_1 + 12t_1{}^2 + 48t_2 - 36t_1 t_2 + 27t_2{}^2)]$	(−2,2)	3
5	Six Hump Camel	$f(t) = (4 - 2.1t_1^2 + \dfrac{t_1^4}{3})t_1^2 + t_1 t_2 + (-4 + 4t_2^2)t_2^2$	(−5,5)	−1.03162
6	Rosenbrock's	$f(t) = \sum_{i=1}^{D-1}\left[100\left(t_{i+1} - t_i^2\right)^2 + \left(t_i - 1\right)^2\right]$	(−100,100)	0
7	Booth	$f(t) = (t_1 + 2t_2 - 7)^2 + (2t_1 + t_2 - 5)^2$	(−10,10)	0
8	Beale	$f(t) = (1.5 - t_1 + t_1 t_2)^2 + (2.25 - t_1 + t_1 t_2^2) + (2.625 - t_1 + t_1 t_2^3)^2$	(−4.5 4.5)	0
9	Schwefel	$f(t) = \sum_{i=1}^{n} \lvert t_i \rvert$	(−2,2)	3
10	Griewank	$f(t) = \sum_{i=1}^{d} \dfrac{t_i^2}{4000} - \prod_{i=1}^{d} \cos\left(\dfrac{t_i}{\sqrt{i}}\right) + 1$	(−100, 100)	0

In order to find the comparative results of IWO with GA, PSO, ABC, and FPA the simulated results have been also analyzed for the better performance. Various benchmark function have been tested on 100 runs with 50 population size. The experimental results are mentioned in Tables 9.1 and 9.2.

TABLE 9.2 Statistical Results of 100 Runs with 30 Iterations Acquired.

Algorithm	Mean	Std	Min	Dim
Easom function				
GA	−1	0.0000	−1	2
PSO	−1	0.0000	0.0000	2
ABC	−1	0.0000	0.0000	2
FPA	0.0000	0.0000	0.0000	2
IWO	0.0000	0.0000	0.0000	2
Matyas function				
GA	0.0000	0.0000	0.0000	2
PSO	0.0000	0.0000	0.0000	2
ABC	0.0000	0.0000	0.0000	2
FPA	0.0000	0.0000	0.0000	2
IWO	0.0000	0.0000	0.0000	2
Rastrigin function				
GA	0.0000	0.0000	51.76753	2
PSO	0.0000	0.0000	41.574323	2

TABLE 9.2 *(Continued)*

Algorithm	Mean	Std	Min	Dim
ABC	0.0000	0.0000	0.0000	2
FPA	0.0000	0.0000	0.0000	2
IWO	0.0000	0.0000	0.0000	2
Goldstien–Price function				
GA	5.250611	5.870093	5.250611	2
PSO	3	0.0000	3	2
ABC	3	0.0000	3	2
FPA	3	0.0000	3	2
IWO	3	0.0000	3	2
Rosenbrock function				
GA	1.96E+05	3.85E+04	1.58342E-06	2
PSO	15.088617	24.170196	15.088617	2
ABC	0.0887707	5.036187	0.0887707	2
FPA	−96.6812	3.3053	−98.9995	2
IWO	−96.6812	3.3053	−98.9995	2
Six hump camel function				
GA	−1.03163	0.0000	−1.03163	2
PSO	−1.0316285	0.0000	−1.0316285	2
ABC	−1.0316285	0.0000	−1.0316285	2
FPA	−1.0316	0.0000	−1.0316	2
IWO	−1.0316	0.0000	−1.0316	2
Booth function				
GA	0.0000	0.0000	0.0000	2
PSO	0.0000	0.0000	0.0000	2
ABC	0.0000	0.0000	0.0000	2
FPA	0.0000	0.0000	0.0000	2
IWO	0.0000	0.0000	0.0000	2
Beale function				
GA	0.0000	0.0000	0.0000	2
PSO	0.0000	0.0000	0.0000	2
ABC	0.0000	0.0000	0.0000	2
FPA	0.0000	0.0000	0.0000	2
IWO	0.0000	0.0000	0.0000	2
Griewank function				
GA	10.63346	1.161455	9.065432	2
PSO	0.01739118	0.020808	0.0123678	2
ABC	0.0000	0.0000	0.0000	2
FPA	0.0015346	0.0027654	0.0011348	2
IWO	0.0000	0.0000	0.0000	2

TABLE 9.2 *(Continued)*

Algorithm	Mean	Std	Min	Dim
Schwefel				
GA	11.0214	1.3868	1.0200	2
PSO	0.0021	0.0018	0.0001	2
ABC	0.0034	0.0031	0.0012	2
FPA	0.0301	0.0121	0.0017	2
IWO	0.0002	0.0001	0.0000	2

Numerical comparison, we plotted the graph between fitness function and number of iteration for 10 benchmark functions (Fig. 9.2).

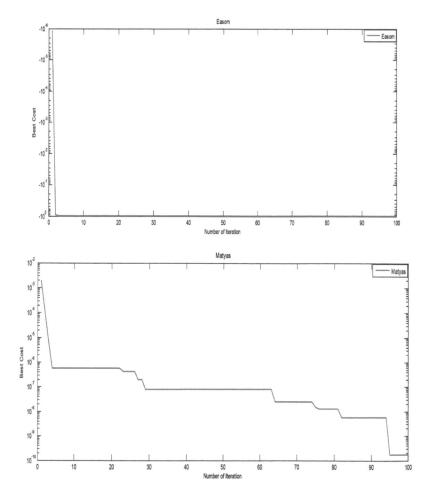

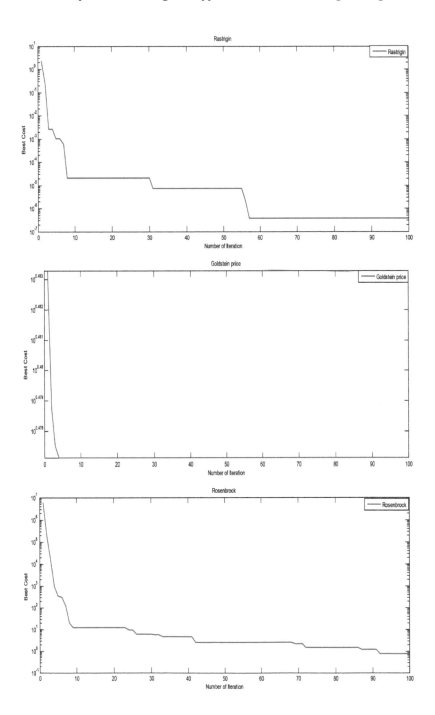

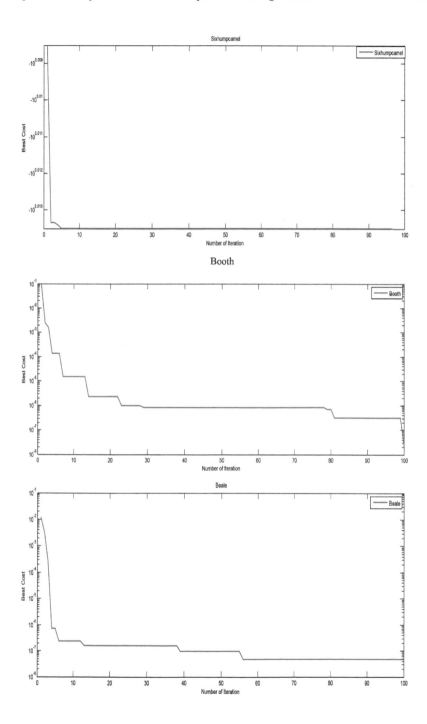

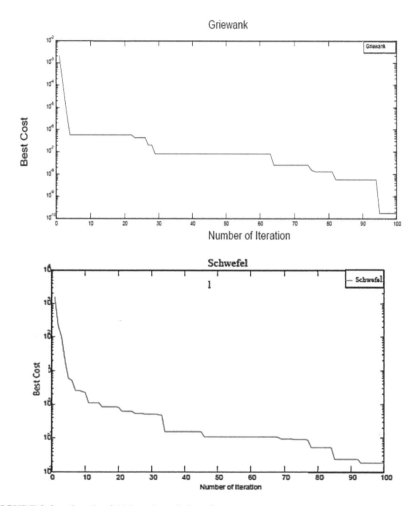

FIGURE 9.2   Graph of 10 benchmark function.

9.3   CONCLUSION

Weeds are known to identify the most suitable environmental conditions for their survival, adapting to the environmental and the other factors and further resisting the changes. Given the behavior of weeds there has been consistent requirement for effective optimization algorithm. One such algorithm includes the IWO. IWO is enthused by the behavior of weed colonies and is a population-based evolutionary optimization method.

IWO efficacy is evaluated via a set of multidimensional benchmark function, in which global and local minima are identified.

In this investigation, we have executed IWO on 10 benchmark functions. This study has made an attempt to carry out the best optimal solution. Further, mean and standard deviations were also obtained. Lastly, statistical comparisons of results were made with other methods, showing that obtained results are more significant and IWO performs better than other algorithms. The study will be useful in contributing to tackle with the threat which weeds poses to agriculture.

KEYWORDS

- **optimization**
- **particle swarm optimization**
- **genetic algorithm**
- **IWO**
- **ABC**
- **FPA**
- **benchmark functions**

REFERENCES

1. Maclaren, C.; Labuschagne, J.; Swanepoel, P. A. Tillage Practices Affect Weeds Differently in Monoculture vs. Crop Rotation. *Soil Tillage Res. (Science Direct)* **2021,** *205,* 1–11.
2. Robertson, G. P.; Grace, P. R. *Greenhouse Gas Fluxes in Tropical and Temperate Agriculture: The Need for a Full Cost Accounting of Global Warming Potentials, Tropical Agriculture in Transition –Opportunities for Mitigating Greenhouse Gas Emissions*; Springer: Dordrecht, 2004; pp 51–63.
3. Moorman, T. B. A Review of Pesticide Effects on Microorganisms and Microbial Processes B Related to Soil Fertility. *J. Product. Agric.* **1989,** 14–23.
4. Hoy, J.; Swanson, N. L.; Seneff, S. The High Cost of Pesticides: Human and Animal Disease. In *Poultry, Fisheries and Wildlife Sciences, Research Article*, 2015; pp 1–19.
5. Morris, A. J.; Bright, J. A.; Winspear, R. J. A Review of Indirect Effects of Pesticides on Birds and Mitigating Land Management Practices. *RSPB Res. Rep.* **2008.**
6. Mehrabian, A. R.; Lucas, C. A Novel Numerical Optimization Algorithm Inspired from Weed Colonization. *Ecol. Inform.* **2006,** *1* (4), 355–366.

7. Zhang, X.; Nui, Y.; Cui, G.; Wang, Y. A Modified Invasive Weed Optimization with Crossover Operation. In *8th World Congress on Intelligent Control and Automation (WCICA)*; IEEE, Jinan, China, 2010; pp 1–11.

8. Chen, H.; Zhou, Y. Q.; He, S. C.; Ouyang, X. X.; Gou, P. G. Invasive Weed Optimization Algorithm for Solving Permutation Flow-Shop Scheduling Problem. *J. Comput. Theor. Nanosci.* **2013,** *10* (3), 708–713.

9. Roy, S.; Islam, S. M.; Das, S.; Ghosh, S.; Vasilakos, A. V. A Simulated Weed Colony System with Sub Regional Differential Evolution for Multimodal Optimization. *Eng. Optim.* **2013,** *45* (4), 459–481.

10. Liu, C.; Wu, H. Synthesis of Thinned Array with Side Lobe Levels Reduction Using Improved Binary Invasive Weed Optimization, *Prog. Electromagn. Res.* **2014,** *37,* 21–30.

11. Barisal, A. K.; Prusty, R. C. Large Scale Economic Dispatch of Power System Using Oppositional Invasive Weed Optimization. *Appl. Soft Comput.* **2015,** *29,* 122–137.

12. Zhao, Y.; Leng, L.; Qian, Z.; Wang, W. A Discrete Hybrid Invasive Weed Optimization Algorithm for the Capacitated Vehicle Routing Problem. *Procedia Comput. Sci.* **2016,** *91,* 978–987.

13. Yaseen, H. T.; Mitras, B. A.; Khidhir, A. S. M. Hybrid Invasive Weed Optimization Algorithm with Chicken Swarm Optimization Algorithm to Solve Global Optimization Problems. *Int. J. Comput. Netw. Commun. Secur.* **2018,** *6* (8), 173–181.

14. Rad, H.; Lucas, C. A Recommended System Based on Invasive Weed Optimization Algorithm. *IEEE Congr. Evol. Comput.* **2007,** 4297–4304.

15. Mallahzadeh, A. R.; Oraizi, H.; Davoodi-Rad, Z. Application of the Invasive Weed Optimization Technique for Antenna Configurations, *Progr. Electromagn. Res.* **2008,** *79,* 137–150.

16. Mallahzadeh, A. R.; Es'haghi, S.; Alipour, A. Design of an E-shaped MIMO Antenna Using IWO Algorithm for Wireless Application at 5.8 GHz. *Progr. Electromagn. Res.* **2009,** *90,* 187–203.

17. Zhang, X.; Wang, Y.; Cui, G.; Niu, Y.; Xu, J. Application of a Novel IWO to the Design of Encoding Sequences for DNA Computing. *Comput. Math. Appl.* **2009,** *57,* 2001–2008.

18. Petropoulos, K. E.; Vrahatis, M. N. Particle Swarm Optimization and Intelligence: Advances and Applications, 2010.

19. Yang, X. S. *Engineering Optimization an Introduction with Meta-Heuristic Applications*; John Wiley & Sons, 2010.

20. Yang, X. S. *Nature-Inspired Meta-Heuristic Algorithms*; Luniver Press, 2010.

21. Karaboga, D. A Comparative Study of Artificial Bee Colony Algorithm. *Turkey Appl. Mathemat. Computat. J.* **2009,** *214,* 108–132.

22. Jatoth, C.; Gangadharan, G. R.; Fiore, U. Optimal Fitness Aware Cloud Service Composition Using Modified Invasive Weed Optimization. *Swarm Evol. Computat. J.* **2019,** *44,* 1073–1091.

23. Uyar, K.; Ulker, E. B – Spline Curve Fitting with Invasive Weed Optimization. *Appl. Mathemat. Modell.* **2017,** *52,* 320–340.

24. Rahmani, Y.; Shahvari, Y.; Kia, F. Application of Invasive Weed Optimization Algorithm for Optimizing the Reloading Pattern of a VVER -1000 Reactor (in Transient Cycles). *Nucl. Eng. Des.* **2021,** *376.*

25. Basak, A.; Maity, D.; Das, S. A Differential Invasive Weed Optimization Algorithm for Improved Global Numerical Optimization. *Appl. Mathemat. Computat.* **2013,** *219,* 6645–6668.
26. Falco, I. D.; Cioppa, A. D.; Maisto, D.; Scafuri, U.; Tarantino, E. Biological Invasion–Inspired Migration in Distributed Evolutionary Algorithms. *Inform. Sci.* **2012,** *207,* 50–65.
27. Ghalenoei, M. R.; Hajimirsadeghi, H.; Lucas, C. Discrete Invasive Weed Optimization Algorithm: Application to Cooperative Multiple Task Assignment of UAVs. In *Joint 48th IEEE Conference on Decision and Control and 28th Chinese Control Conference*; Shanghai, 2009; pp 1665–1670.
28. Giri, R.; Chowdhury, A.; Ghosh, A.; Das, S.; Abraham, A.; Snasel, V. A Modified Invasive Weed Optimization Algorithm for Training of Feed-Forward Neural Networks. In *IEEE International Conference on Systems*; Istanbul, 2010; pp 3166–3173.
29. Jose, S.; Singh, H. P.; Batish, D. R.; Kohli, R. K. *Invasive Plant Ecology*; Taylor & Francis Group, LLC: Boca Raton, 2013.
30. Kundu, D.; Suresh, K.; Ghosh, S.; Das, S.; Panigrahi, B. K.; Das, S. Multi-objective Optimization with Artificial Weed Colonies. *Inform. Sci.* **2012,** *181,* 2441–2454.
31. Li, Y.; Yang, F.; Ouyang, J.; Zhou, H. Yagi-Uda Antenna Optimization Based on Invasive Weed Optimization Method. *Electromagnetics* **2011,** *31,* 571–577.

CHAPTER 10

An Overview of Computational Tools

NAVNEET KAUR*, SHALINI SAHAY, and SHRUTI K. DIXIT

Electronics and Communication Department, Rajiv Gandhi, Proudyogiki Vishwavidyalaya (RGPV), Sagar Institute of Research and Technology, Bhopal, Madhya Pradesh, India

Corresponding author. E-mail: navec2000@gmail.com

ABSTRACT

Computational techniques is a set of bio-inspired computational methodologies and tactics to initiate complex real-world problems for which traditional simulation methods cannot be very useful. The methods are very close to the human's way of reasoning and it is able to produce control actions in an adaptive way. This chapter explores the use of computational intelligence techniques (specifically fuzzy logic, artificial neural network and particle swarm optimization) for classification of data. The artificial intelligence mainly includes learning, reasoning, and perception used across different industries. Fuzzy logic sets enable to quantify knowledge. In ANN, the progression of the brain is used to implement algorithms that aim to develop complex design problems. Fuzzy logic is one of the disciplines in artificial intelligence which emulates human reasoning in terms of linguistic variables. There is similarity between fuzzy logic and neural networks. These methods are used to resolve problems in which any mathematical model is not involved. Systems which are implemented using combination of both fuzzy logic and neural networks are termed as neuro-fuzzy systems. These hybrid systems combine advantages of both fuzzy logic and neural networks to perform in an efficient manner.

Computational Intelligence Applications for Software Engineering Problems. Parma Nand, PhD, Rakesh Nitin, PhD, Arun Prakash Agrawal, PhD & Vishal Jain, PhD (Eds.)
© 2023 Apple Academic Press, Inc. Co-published with CRC Press (Taylor & Francis)

Particle swarm optimization (PSO) is a population-based optimization technique which inspired by the motion of bird flocks and schooling fish and many more nature inspired algorithms. The choices of its settings like position, social factors, least and highest velocity plays a vital role in PSO performance. Fuzzy logic is used to select these values. Fuzzy logic-based PSO is implemented to compute settings for the entire swarm.

The other hybrid techniques can also be implemented to enhance the applications. These techniques are the combinations of all computational methods.

10.1   INTRODUCTION

The real-time adaptation and self-organization concepts, paradigms, algorithms, and implementations which assist or expedite appropriate actions in complicated and varying environments are included by Computational Intelligence (CI). It is a collection of bioinspired computational approaches and tactics to initiate complex real-world problems for which traditional simulation methods are not advantageous for definite causes: the steps can be too complex for precise reasoning, which contains some ambiguities which can be of stochastic class. Quite a few real-life problems cannot be translated into digital for machines to process it in real world. The human's way of reasoning and ability to produce control actions in an adaptive way results in inexact and incomplete knowledge which can be used by these methods. Therefore, artificial intelligence is a combination of mainly three techniques. The fuzzy logic which follows the natural language, the neural networks which initiate to learn empirical data, and evolutionary computing techniques based on the progression of bioselection, learning theory, and, probabilistic methods which deals with indecision imprecision. This chapter explores the use of computational intelligence techniques (specifically fuzzy logic, artificial neural network, and particle swarm optimization) for classification of data. The artificial intelligence mainly includes learning, reasoning, and perception used across different industries. Fuzzy logic sets enable to quantify knowledge, intuition to model complex problems, was introduced by Zadeh (1965). Particle swarm optimization (PSO) is an optimization technique which is inspired by the motion of bird flocks, schooling fish, and other nature inspired algorithms. The potential solutions take actions in the problem space by following the existing optimum particles in PSO called particles. Section 10.2 describes the ANN, Section 10.3 discusses the fuzzy logic system, and PSO is detailed in Section 10.4.

10.2   ARTIFICIAL NEURAL NETWORK

The biological neurons which activate under specific circumstances results by an action performed by the body are inspired from neural networks. One of the important tools and approaches used in machine learning algorithms is the neural network. For switching those in ON/OFF[1] mode, various layers of interconnected nets powered by activation functions help them. In the training phase similar to other traditional machine algorithms there are also certain values that neural nets learn. These networks are represented by deep learning using artificial intelligence, being applied in many real-time problems, including speech and image recognition, spam email filtering, finance, and medical diagnosis and many more. These algorithms generally do not require any specific rules. For many number of examples that have been implemented, starts to process new, unseen inputs and return successfully accurate results.[2,3]

Multiplied values of inputs and random weights are added with each neuron which is then summed up with static bias value (unique to each neuron layer), then the final value is determined after passing through an appropriate activation function. The weights are adjusted to make the minimum loss by using an error function (input vs. output) is calculated and backpropagation is implemented.[4,5] Figure 10.1 represents the elementary structure of neural network.

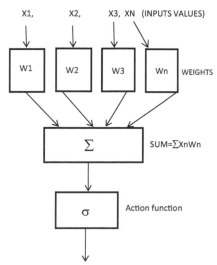

FIGURE 10.1   The elementary structure of neural network.[1]

Source: Reprinted with permission from Ref. [1]. © 2005 John Wiley & Sons, Ltd.

Weights represent numerical values which are multiplied with inputs, and then they are modified and rearranged to reduce the loss in backpropagation. Based on the difference between predicted outputs vs. training inputs they self-adjust themselves in a correct way.[6,7] Then neurons are being switched between ON/OFF through activation function which is a type of mathematical formula. Figure 10.2 shows the layers of neural network.

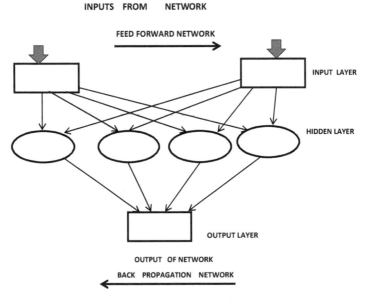

FIGURE 10.2   The various layers of neural network.[1]

Source: Reprinted with permission from Ref. [1]. © 2005 John Wiley & Sons, Ltd.

- The input vector indicates the components of input layer.
- The transitional nodes which distribute the input space into regions with lenient boundaries are shown by hidden layer.
- The output of the final network represented by the output layer.

10.2.1   NEURAL NETWORK TOPOLOGIES

The following are the major neural topologies.

10.2.1.1   PERCEPTRON (P)

A single-layer neural network is known as the perceptron model, which comprises of exclusively two layers: Input Layer, Output Layer, and no

Hidden layers.[8,9] For each node , it estimates the weight for the input, which is shown in Figure 10.3.Then it uses a sigmoid function for activation ,which is further used for classification purposes.

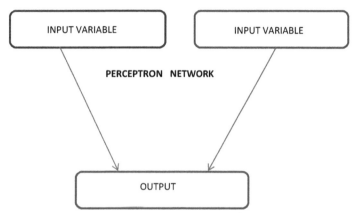

FIGURE 10.3   The illustration of the perceptron model.[3]

➤ Advantages of Perceptron
   They can implement AND, OR, NOR, and NAND gates.
➤ Disadvantages of Perceptron
   Linearly separable problems such as Boolean AND can be implemented by it and does not work for nonlinear problems such as Boolean XOR problem.
➤ Applications
   • Used in Classification of problems
   • Applied in Encode Database (Multilayer Perceptron).
   • For Monitor Access Data (Multilayer Perceptron).

10.2.1.2   FEEDFORWARD (FF) NEURAL NETWORK

It is a network in which the nodes do not form a cycle. All of the inputs are arranged in layers and the output is shown by the output layer. There is no connection between hidden layers with the outer world. All nodes are connected with each other. In general the backpropagation algorithm is updated and the weight values are so adjusted which minimize the prediction error, as shown in Figure 10.4.

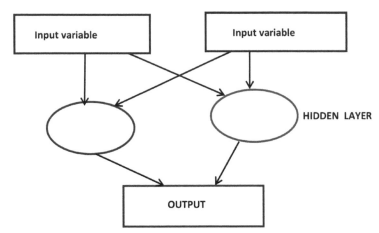

FIGURE 10.4   Demonstration of a feedforward neural network.[2]

➤ Advantages of feedforward neural networks
  • Easy to design and maintain as simple in architecture
  • Very fast execution
  • Highly sensitive for noisy data
➤ Disadvantages of feedforward neural networks
  • Due to the absence of dense layers and backpropagation cannot be used for deep learning.
➤ Applications of feedforward neural networks:
  • Used in Data Compression.
  • Applied in Pattern Recognition.
  • Depleted in Computer Vision.
  • Deployed in Sonar Target Recognition.

10.2.1.3 MULTILAYER PERCEPTRON

As independent node is connected with all neurons in the preceding layers and makes it a distinct neural network. It comprises of bidirectional propagation, that is, forward and backward propagation.[10]

The weights are multiplied by the inputs are fed to the activation function, which is based on backpropagation, reduces the error. For output layer activation function like the nonlinear functions are applied.

➤ Advantages of Multilayer Perceptron
  Expended for deep learning

> Disadvantages of Multilayer Perceptron
  • Little complex to design
> Applications of Multilayer Perceptron
  • Used in Speech Recognition
  • For Machine Translation

10.2.1.4   RADIAL BASIS NETWORK (RBN)

These kinds of networks are mostly employed in function approximation problems. To find whether the answer is an yes or a no, a logical function (sigmoid function) produces an output between zero and one. The RBIs yield the resulted output from the target output, which is very advantageous in case of persistent values. Figure 10.5 shows the radial basis network (RBN).

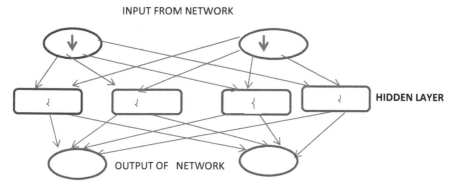

FIGURE 10.5   Illustration of a radial basis network.[4]

> Advantages
  • The training phase is faster as there is no backpropagation learning involved.
  • The roles of Hidden layer nodes are very easy.
> Disadvantage
  Classification is slow in comparison to Multilayer Perceptron due to fact that every node in hidden layer has to compute the RBF function for the input sample vector during classification.

➤   Applications
   • Used in Function Approximation.
   • For Time series Prediction.
   • In Classification and System Control.

10.2.1.5   KOHONEN SELF-ORGANIZING NEURAL NETWORK

A distinct chart consists of neurons which are trained to form its own organization of the training data, is used in the Kohonen map. The input vectors
of arbitrary dimensions are mapped. Either one or two dimensions are
mapped by it. The zone of the neuron remains constant but the weights are
changing subjected on the value during the training. The process involves
various parts, in the beginning state; all neurons begin with a small weight
and the input vector. The nearest one to the point is the "winning neuron"
and the neurons closet to the winning neuron will also moves closes to the
point like in the graphic as shown in Figure 10.6, for the second phase.
The Euclidean distance is the distance between the point and the neurons
is estimated with the minimum distance to wins. Total points are clustered
through the iterations and neuron represents a unique cluster uniquely,
which is the main objective of the organization map.

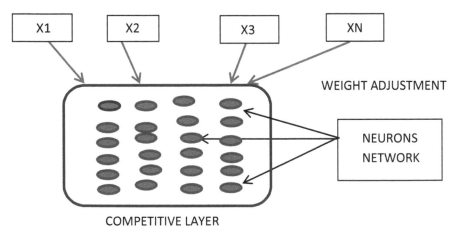

FIGURE 10.6   Representation of a Kohonen neural network.[7]

10.2.1.6   CONVOLUTIONAL NEURAL NETWORK

The features are taken in batch-wise similar to a filter. In this neural network, the network memorizes the parts of the images, which can be implemented during process. The translation of the image from RGB or HSI scale to the Gray scale is computed. The deviations in the pixel value will find out the edges and images in different categories as shown in Figure 10.7. They are brought in a world of difference for introduction of marketing and advertising of the data. These networks are implemented in signal processing and image classification techniques.

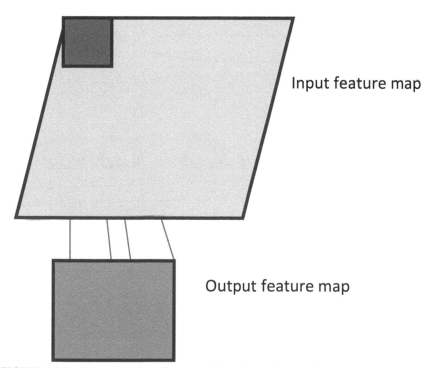

Input feature map

Output feature map

FIGURE 10.7   Representation of a convolutional neural network.

➤ Applications:
   • In Decoding Facial Recognition
   • For Analyzing Documents
   • In Historic and Environmental Collections

10.2.1.7 MODULAR NEURAL NETWORK

Assemblies of different networks working independently and establishing the outputs are formed by these networks. Inputs that are exclusive are compared with other networks which are constructed to perform the subtasks. The networks do not mix with each other in compiling the tasks. Figure 10.8 shows the elementary structure of the network.

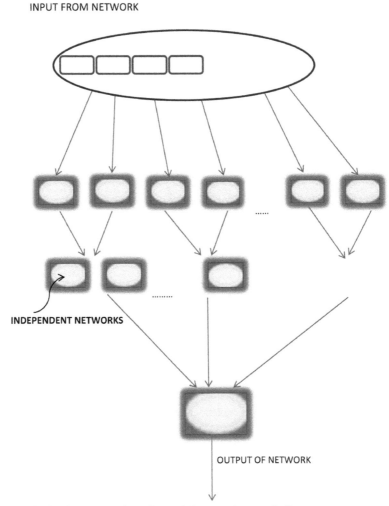

FIGURE 10.8   Representation of a modular neural networks.[7]

➤ Advantages of Modular Neural Network
  1. Very competent
  2. Training and Robustness works independently
➤ Disadvantages of Modular Neural Network
  • With the movement of target problems reduces the speed.
➤ Applications:
  • In Machine Translation and Robot Control.
  • For Prediction in time series anomaly Detection
  • Recognition and Synthesis of speech.
  • Learning of music composition rhythmically.

10.2.1.8   RECURRENT NEURAL NETWORK

For a specific delay in time each neurons in hidden layers receives an input. There is no increase in the dimension of the input, and in the computations of the model are based on the previous historical information. The slow computational speed is the major drawback of this network and is not consider for prediction of any future input of the present state, which is shown in Figure 10.9.

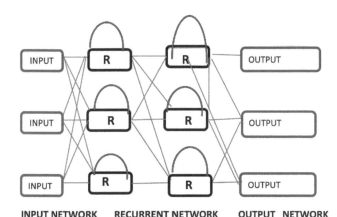

INPUT NETWORK   RECURRENT NETWORK   OUTPUT NETWORK

FIGURE 10.9   Representation of recurrent neural network (RNN).[8]

➤ Advantages of RNNs are
  • Used with model sequential data
  • The convolution layers are used to extend the pixel effectiveness.

➤ Disadvantages of RNNs
  • Exploding problems like Gradient vanishing
  • A difficult task is training the network
  • Very difficult to process elongated sequential data.

10.2.2   APPLICATIONS OF NEURAL NETWORKS

A broad range of situations can assess various inputs, including images, videos, files, databases, and many more. Any implicit programming is not required to interpret the inputs. There is no limit to the areas of virtual applications due to the generalized approach of neural networks. There are enormous neural nets applications increasing day by day in medicine, normally they are linked to diagnostics systems. However, they are not only able to identify examples, but sustain very important information. The prominent applications involve the interpretation of medical data.

10.3   FUZZY LOGIC SYSTEM

A multivalued logic form in which variables are assigned truth values between 0 and 1 is termed as fuzzy logic. In some situations in real life, the statement cannot be determined as true or false. In this case, flexibility for reasoning is provided by fuzzy logic. The fuzzy logic method follows the human style of decision-making which indicates all values between digital values T and F. Fuzzy algorithm is applicable in various fields, from control theory to Artificial Intelligence.

10.3.1   CHARACTERISTICS OF FUZZY LOGIC

Principal characteristics of fuzzy logic are:

  • Flexible technique of machine learning and implementation is easy.
  • Make easier to imitate the logic of human thought
  • Two possible solutions are represented according to two values of logic.
  • Provides fairly accurate reasoning.
  • Nonlinear functions of arbitrary complexity are built.

10.3.2   *FUZZY LOGIC ARCHITECTURE*[11]

There are four main parts in fuzzy logic architecture, as shown in Figure 10.10.

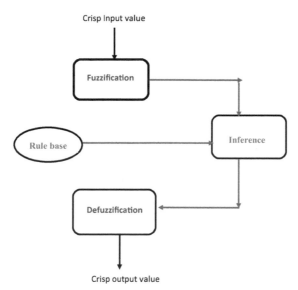

FIGURE 10.10   Fuzzy logic architecture.

- **Rule Base:**

It consists of rules formed by using IF–THEN conditions to regulate the decision-making system. The quantity of the fuzzy set of rules is reduced.

- **Fuzzification:**

Fuzzification is used to translate inputs. The crisp numbers are converted into fuzzy sets. The Fuzzification section is utilized to transform inputs as well as in dividing the input signals into the following five categories of membership functions:

  – Large positive value (LP).
  – Medium positive value (MP).
  – Small value (s).
  – Medium negative value (MN).
  – Large negative value (LN).

- **Inference Engine:**

The extent of correspondence is determined between fuzzy input and fuzzy rules using this component. The rule which is required to be implemented is decided based on that proportion. Then, the control actions are developed by combining the applied rules. This practice helps in creating the process of human thinking by applying fuzzy implication and IF–THEN rules on the inputs.

- **Defuzzification:**

The reverse process of the Fuzzifier is performed here. The fuzzy values are translated to the conventional numerical signals and utilized to control the system operation.

The defuzzification process is performed in the final stage to transform the set of fuzzy values into a crisp value. There are a number of methods for fuzzification, and the programmer selects the appropriate method according to the requirement.

10.3.2.1   *LINGUISTIC VARIABLES*

The input variables and output variables of the system are represented by linguistic variables. These variables do not consist of numbers but include words from the languages. A set of linguistic terms are formed from linguistic variables.

10.3.2.2   *MEMBERSHIP FUNCTIONS*

The linguistic terms are quantified using membership functions and fuzzy sets are represented in the graphical form. Suppose there is a fuzzy set X and Y represents the set from which value of membership function is taken, then a membership function is defined as $\mu_x: Y \to [0,1]$.

Here, each entity of Y is assigned a value between 0 and 1. This is termed as degree of membership. In graphical form,

- x-axis characterizes the universal set.
- y-axis denotes the membership degree in the interval $[0,1]$.

Several membership functions are valid for fuzzifying a numerical value. Membership functions are shown in Figure 10.11.

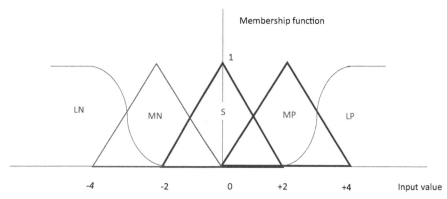

FIGURE 10.11   Membership functions.[11]

Various membership functions are used such as Triangular, Piecewise linear, singleton, trapezoidal, Gaussian but straight line functions like triangular and trapezoidal membership function shapes are most common.

10.3.2.3   FUZZY LOGIC CONTROLLER (FLC)

A fuzzy logic controller carries out the operation of providing the linguistic-information-based outputs. It achieves fairly accurate reasoning based on the human interpretation to achieve control logic. The controller comprises of the units knowledge base as well as inference engine. The fuzzy rules and the membership functions are part of knowledge base, which requires knowledge of the system operation.[12]

10.3.3   BENEFITS AND DRAWBACKS OF FUZZY LOGIC SYSTEM

Benefits of Fuzzy Logic System (FLS)

- Mathematical concepts used in fuzzy analysis are minimal.
- Addition or deletion of rules can modify FLS and thus provides flexibility.
- Indeterminate input information can be given to FLS.

- Fuzzy logic systems can be easily created and understood.
- Fuzzy logic can resolve complicated real-life problems, including medicine, as it approximates human perceptive and decisiveness.
- The structure of FLS is explicable
- The common use of fuzzy logic is for industrial and practicable purposes
- The machines and other customer products can be controlled by using fuzzy logic in soft computing.
- The system is robust because no specific inputs are needed.
- The system performance can be improved by easy modifications.
- Use of inexpensive sensors reduces the complexity as well as system cost.

Drawbacks of FLS

- Fuzzy system designing does not possess systematic approach.
- The simple approach only makes it understandable.
- They are appropriately used for the problems where accuracy is not needed.
- Fuzzy logic does not provide accurate results always. The solutions are established on assumption, thus not extensively accepted.
- Testing with hardware is required for verification of a knowledge-based system.
- The setting of rules using IF–THEN and the formation of membership functions is a difficult task

10.3.4   IMPLEMENTATION OF FUZZY LOGIC SYSTEM[13]

Fuzzy logic can be employed in systems with varying range and capacity. It can also be implemented in Artificial Intelligence, in hardware or software, or combination of both.

How is fuzzy logic used?

1. The user should be precise about the problem, the desired outcome, the probable system failures and methods to control the system.
2. The correlation between input and output should be determined. The lowest possible number of variables should be chosen for input to system.

3. Membership function should be created for defining range of values of input and output variables.

4. The problem should be formulated into a sequence of "IF AND THEN" rules by using structure of fuzzy logic. Rules should be framed based on required outcome for given conditions.

5. Different operations on fuzzy set execute valuation of rules. Evaluate rules in the rule base.

6. The results of evaluation are combined to generate a final outcome. This outcome is a fuzzy value.

7. The output data is converted into crisp values. Defuzzification is implemented corresponding to membership function used for output variable.

8. System testing and evaluating results regulate the membership functions and rules.

10.3.5 FUZZY LOGIC APPLICATIONS

Fuzzy logic systems is applicable in

- Engineering fields where there is improbability.
- Fields where appropriate reasoning is acceptable if accurate reasoning is not possible.

The unique features of fuzzy logic make it an appropriate alternative for solving various problems:

1. The robust property of fuzzy system does not require noise-free inputs and it can be designed to work safely in case of failure of a feedback sensor.

2. FLC manages rules defined by users which can be easily modified to enhance performance of the system. New sensors can be designed by creating appropriate rules in the system.

3. The operation of system is rule based so require number of inputs and outputs can be generated. After defining rules, the interrelations between input and output variables must be defined. The control system can be divided into smaller sections and numerous fuzzy logic controllers are developed to control these sections.

4. The nonlinear systems that are difficult to be accomplished mathematically can be managed by fuzzy system.

These are some applications of fuzzy system.
a. Automotive Systems
    - Vehicle environment control
    - Automatic Gearboxes
b. Consumer Electronic Goods
    - Video Cameras
    - Photocopiers
    - Television
c. Environment Control
    - Air Conditioners
    - Heaters
    - Humidifiers
    - Dryers
d. Domestic Goods
    - Vacuum Cleaners
    - Washing Machines
    - Microwave Ovens
    - Refrigerators

Fuzzy logic is also used in diverse sectors mentioned below:

In Industries: The approach used in fuzzy control is characterized by control rules. The applications include cement kiln control, smart engine control, mechanized train systems, proficiency optimization in induction motor, water quality control, etc.

Products: Fuzzy system is implemented in user goods like heaters and television sets. Recent work is aimed at developing handy products based on fuzzy logic.

There is a vast scope of fuzzy-logic-based products in the recent times. The fuzzy logic system finds application in mobile communication system. Fuzzy logic system has become apparent as a dominant system to handle complicated problems. It contributes to manage improbability of complicated engineering problems. The problems that cannot be resolved through conventional numerical methods are managed by employing fuzzy logic approach. In the situations where classical control methods are available, fuzzy logic is introduced to improve the controller performance and also to simplify the control algorithm.

Some authors[14,15] implemented adaptive fuzzy logic controllers which used inference parameters to execute adaptive fuzzy inference methods.

The membership functions are selected depending on human practice which should have significance for fuzzy controllers design.

In dynamic networks like MANETs, fuzzy logic and genetic algorithm can be effectively used to improve decision-making, and to optimize protocol parameters with dynamism.

Fuzzy logic system is applied with recent research areas in MANET, signal processing, and communication systems.[16–18] Though, it is a sophisticated method of decision-making based on several criteria.[19,20] Route selection process is a significant problem for the nodes in MANETs. Selection of optimum routes using fuzzy logic results in improved performance of MANET. The fuzzy-based route selection can improve the performance parameters like end-to-end delay, energy consumption, and throughput.

MANET nodes are power operated and function with restricted battery power and affect the life span of the mobile node. The communication through mobile nodes should be done in a wise manner. In MANET, accurate route is selected using fuzzy logic and parameters such as energy of nodes, network bandwidth, shortest distance between nodes and route stability. The route setup time required is reduced.

Currently, many researches implemented a fuzzy approach to resolve crucial issues in MANET. Chaudhary et al.[21] implemented a protocol where nodes can participate in route establishment formulated on the approach using fuzzy logic. Anuradha and Mala[22] developed a routing protocol to detect and control congestion based on cross-layer approach through fuzzy inference rules. Robinson et al.[23] developed a technique using fuzzy-logic-based mutual sharing methodology which aims at reducing the detection time of misbehaving nodes. An efficient routing is provided by controlling security attacks due to black holes.

10.4 PARTICLE SWARM OPTIMIZATION

Optimization mechanisms dependent on natural algorithm have been turned out to be very powerful and transformative. These algorithms show versatile, powerful, and compelling conduct as nature does.[24,25] A versatile implies that it enhances its objective of accomplishing capability after some time; vigorous means it is adaptable and never separates while successful indicates that it obtains an acceptable solution. It fundamentally endeavors to join essential heuristic techniques in advanced stage frameworks intended at productively and viably investigating a search space. These types of algorithms are known as meta-heuristics. Under the category of meta-heuristics, Ant Colony Optimization (ACO),[26] Artificial Neural Network (ANN), Evolutionary Computation (EC) methods, fuzzy sets; Genetic Algorithms (GA), PSO, etc. are assortments of heuristic procedures.[27] A heuristic search technique PSO builds on the concept of collaborative actions and swarming in biological populations. The behavior of social animals like flock of birds, swarm of honeybees, or school of fishes, etc., motivates us to develop the optimization techniques.[28] Adaptableness is the main advantage of this heuristic method. It is computationally more efficient (uses less number of function evaluations) and requires less time. The velocity and position updating in PSO is dependent on easy equations; it is effectively utilized on large data sets. The accurate solutions are generated by PSO using less number of iterations.

The social behavior of animals and birds inspires the development of stochastic optimization technique known as PSO. It was first presented in 1945 by Kennedy and Eberhart (1995) as far as social and cognitive behavior is concern. The different areas for instance engineering, power systems, data mining, etc., problems are resolved by this technique.

PSO manages issues in which the best solution can be addressed as a point or surface in an n-dimensional space. In this space, hypotheses are plotted and developed with an initial velocity, just like an establishment of a communication channel in between the particles. After every time step, particles by then pass through the arrangement space and are evaluated by some health measure. After some time, the particles are quickened toward those particles inside their correspondence gathering having excellent fitness values. The primary favorable position of a methodology over other worldwide minimization methodologies, for example, simulated annealing is the huge quantities of individuals that make up the molecule swarm make the strategy amazingly versatile to the difficulty of local minima. PSO is a meta-heuristic technique as it makes very few or no assumptions about the issue being improved. It can glance through extraordinarily huge spaces of candidate solutions. In any case, meta-heuristics, for example, PSO do not guarantee an optimal solution is ever found. Likewise, PSO does not utilize the gradient of the problem being improved, which implies PSO does not need that the advancement issue be differentiable as is needed by exemplary optimization techniques for example quasi-newton technique. Each single arrangement is a bird' in the search space (particle). The value to be optimized is calculated by the fitness function and combined with each particle. The flying of the particle expresses that it has the velocity. In the search space, the particles fly and fiddle with the velocities dynamically as per their previous activities. This procedure guides the particles to fly in the direction of the improved search area. The swarm intelligence (SI)[29] concept is appropriate for getting solution to the problem. It is generally a novel bioinspired gathering strategy searches for inspiration behavior of flock of birds, swarm of insects or other animals. PSO is simple to implement and maintain a kind of memory which is necessary to get an optimal solution. At the beginning, the nodes are perceived by imparting a path request signal. In the wake of getting route requests, any hub indistinguishable to swarm particle instated speed and position uses the possibility of algorithm. Swarms ramble through a 3D space and looking forward to the most ideal path. Particles are viewed as individuals and the whole population is seen as a swarm. Swarms are outlined first and the quantity of occupants in quite a while is apportioned self-emphatically over the search space. The interaction is rehashed and once more; every molecule is rearranged by two "best" values, singular best past area called pBest and swarm best past area gBest. Stepwise calculations are summed up as given below:

a. Individuals are designated by discretionary position at first.
b. The best worth is approximated for individual.
c. The person's best worth is contrasted and its pBest for every individual particle. This pbest is contrasted with the current particle's position. Updation of pBest is conceivable in the event that the current particle's position is prevalent than the pBest value, then pBest is set with the current particle's position, x as p.
d. The individual having the best wellness worth will be perceived as gBest wellness work.
e. Alteration of pBest and gBest of all the individuals using (1) and (2).
f. Reiterating advance 2{5 until a satisfactory worth is achieved}.

The advancement of the individual is coordinated by reestablishing its speed and position boundaries as indicated by following condition:-

$$V_i^{t+1} = wV_i^t + c_1 r_1(pbest(t) - X_i^t) + c_2 r_2(gbest(t) - X_i^t) \qquad (10.1)$$

$$X_i^{t+1} = X_i^t + V_i^{t+1} \qquad (10.2)$$

c—Acceleration coefficient, w—Inertia weight, r—the random values between 0 and 1,

X—Position of particle, V—Velocity of particle.

10.4.1 SELECTION OF PARAMETERS

PSO parameters decision affects optimization performance. PSO parameters selection that yields extraordinary execution has been the subject of numerous assessments. The PSO parameters can in similar manner be tuned by using another overlaying enhancer, a thought is known as meta-optimization or even tweaked during the advancement, for example, by methods for fuzzy logic. Parameters have furthermore been tuned for various smoothing out circumstances.[30]

10.4.2 NEIGHBORHOOD AND TOPOLOGIES

The subset of particles defined by the topology of the swarm with which each particle can exchange information. The global topology as the swarm communication structure is utilized by the fundamental version of the algorithm. This topology allows all particles to establish communication with all the other particles. In this way, the whole swarm shares the same best position from a single particle. Regardless, this technique may lead the swarm to be caught into a local minimum. Hence different geographies have been utilized to control the progression of data among particles. In neighborhood geographies, particles just discuss data with a subset of particles. The subset can be a mathematical one for example, "them closest particles"—or, all the more regularly, a social one, for example, a bunch of particles that is autonomous of distance. In such cases, the PSO variety should be neighborhood best (versus overall best for the key PSO).

10.4.3 HYBRIDIZATION

PSO was united with some traditional and transformative optimization algorithms to take the potential gains of the two strategies and reimburse the weaknesses of each other. This sort of PSO is called hybridized PSO. Few examples of hybridization techniques are: (1) GA, (2) Tabu search (TS), (3) Simulated annealing (SA), (4) ACO, (5). Biogeography-based optimization (BBO), (6) Artificial Immune System (AIS) and (7) Harmonic search (HS)

1. Roused by Charles Darwin's hypothesis of common advancement, **GA** is a search heuristic. This calculation mirrors the interaction of characteristic determination in which the fittest individuals are chosen for proliferation to deliver offspring of the future.

2. TS method is used to solve combinatorial optimization problems (an optimal ordering and selection of options is desired). The areas of resource planning, financial analysis, scheduling, environmental conservation, space planning, telecommunications, molecular engineering, flexible manufacturing, waste management, biomedical analysis, pattern classification, logistics, energy distribution, and scores of others are the applications of TS method.

3. **SA** is a strategy based on probabilistic approximating the global optimum of a given function. It is a meta-heuristic optimization method based on large search space. It is regularly utilized when the inquiry space is discrete (e.g., the traveling salesman problem).

4. **ACO** is a probabilistic strategy for taking care of computational issues which can be decreased to discover great ways through diagrams. Artificial ants address for multispecialist strategies energized by the conduct of genuine ants. The communication between natural ants is pheromone based and is routinely the overwhelming worldview utilized. An amalgamation of artificial ants as well as neighborhood search calculations have become a method for different enhancement assignments including a category of chart, for example, vehicle and internet routing.

5. **BBO** is a developmental calculation that upgrades a capacity by stochastically and iteratively improving up-and-comer arrangements concerning a given proportion of value, or wellness work. BBO has a place with the class of meta-heuristics since it incorporates numerous varieties, and since it does not make any presumptions about the issue and can in this manner be applied to a wide class of issues.

6. The discipline of **AIS** is all about abstracting the construction and capacity of the invulnerable framework for computational frameworks. It examines the use of the frameworks toward tackling computational issues in the field of math, data innovation as well as engineering. It is a subfield of biologically propelled enrolling

and Natural estimation, with interests in Machine Learning and having a spot with the more broad area of Artificial Intelligence. The discipline has adaptable structures, spurred by hypothetical immunology and saw safe limits. The principles and models are applied to give solution for problems.

7. **HS** method is a technique that is enlivened by the basic standards of the artists' extemporization of the amicability. HS has covered various zones including industry, advancement benchmarks, power frameworks, clinical science, control frameworks, development plan, and data innovation

10.4.4   BENEFITS AND DRAWBACKS OF THE PSO ALGORITHM

Benefits of the algorithm are summarized as: very easy concept, have not many parameters to change, ready to run parallel computation, have short computational time, simple implementation, strength to control boundaries, and high computational competence.

Drawbacks: (1) It is difficult to characterize initial design parameters. (2) It can combine prematurely and be caught into a nearby least especially with complex issues. (3). It cannot work out the issues of scattering.

10.4.5   APPLICATIONS AREAS OF PSO ALGORITHM[31,32]

1. Engineering: Electrical and Electronics, Automatic, Chemical, Civil, Mechanical, Biological, etc.
2. Medicine
3. Fuel as well as energy
4. Operations
5. Communications

10.5   CONCLUSION

Computational techniques are the process in which computers are used to investigate and resolve technical problems. Due to exponential growth in the field of computers and the data volume, it is required to develop new methods and techniques necessary for information which cannot be detected by traditional means employing human capacity of analysis. Computational techniques are reliable and effective method for solving statistical, technical, engineering, ecological, and numerical problems by using computers. Therefore, the step-wise approach is used for resolving problems in computational technique. In all Engineering Sciences applications, computational technique is used to tackle design problems and enhance the performance of the system. The development of computational methods has assisted in the progression of complex systems and treating simulation problems. The different Engineering fields have benefited from the advancement of computational techniques. This chapter discussed about various computational techniques like ANN, PSO, and fuzzy logic system.

In ANN, the progression of the brain is used to implement algorithms that aim to develop complex design problems. Fuzzy logic is one of the disciplines in artificial intelligence which emulates human reasoning in terms of linguistic variables. There is a similarity between fuzzy logic and neural networks. These methods are used to resolve problems in which any mathematical model is not involved. Systems that are implemented using combination of both fuzzy logic and neural networks are termed as neuro-fuzzy systems. These hybrid systems combine advantages of both fuzzy logic and neural networks to perform in an efficient manner.

PSO is a computational method for optimization by using repetitive approach and optimize solution of candidate with regard to a specified degree of quality. The population of particles is moved around with given particle's position and velocity in accordance with simple arithmetic formulae. The choices of its settings like position, social factors, least and highest velocity plays a vital role in PSO performance. Fuzzy logic is used to select these values. Fuzzy-logic-based PSO is implemented to compute settings for the entire swarm.

The other hybrid techniques can also be implemented to enhance the applications. These techniques are the combinations of all computational methods.

KEYWORDS

- **computational tools**
- **artificial neural network**
- **fuzzy logic system**
- **particle swarm optimization**
- **optimization techniques**

REFERENCES

1. Abraham, A. Artificial Neural Networks. In *Stillwater*; OK, USA, 2005.
2. Kumar, J. G. Artificial Neural Network and its Application; IARI: New Delhi.
3. Nihal, P. et al. Instant E-Learning, a Chat Room Concept for Education. *Int. J. Eng. Innov. Res.* **2012,** *4* (6), 820–822.
4. Gaeta, M.; Miranda, S.; Orciuoli, F.; Paolozzi, S.; Poce, A. An Approach to Personalized E-Learning, Systemics. *Cybern. Info.* **2013,** *11* (1).
5. Ahmed, R. K. A. Artificial Neural Networks in E-Learning Personalization: A Review. *Int. J. Intell. Info. Syst.* **2016,** *5* (6), 104–108. DOI: 10.11648/j.ijiis.20160506.14, pp. 2328–7675.
6. Baylari, A.; Montazer, G. A. Design a Personalized E-Learning System Based on Item Response Theory and Artificial Neural Network Approach. *Expert Syst. App.* **2009,** *36*, 8013–8021. http://dx.doi.org/10.1016/j.eswa.2008.10.080.
7. Lee, K. Y.; Chung, N.; Hwang, S. Application of an Artificial Neural Network (ANN) Model for Predicting Mosquito Abundances in Urban Areas. *Ecol. Info.* **2016,** *36*, 172–180.
8. Moon, S. W.; Kong, S. G. Block-Based Neural Networks. *IEEE Trans. Neural Netw.* **2001,** *12* (2), 307–317.
9. Kolekar, S. V. Learning Style Recognition Using Artificial Neural Network for Adaptive User Interface in E-learning, 2010. IEEE, ISBN 978-1-4244-5967-4/10/$26.00.
10. Nagendra, P.; Dey, S. H. N.; Dutta, T. Artificial Neural Network Application for Power Transfer Capability and Voltage Calculations in Multi-Area Power System. *Leonardo Electron. J. Pract. Technol.* Jan–June **2010,** *16*, 119–128. ISSN: 1583-1078.
11. Zadeh, L. A. Communication Fuzzy Algorithms. *Info. Control* **1968,** *12*, 94–102.
12. Dadios, E. P. *Fuzzy Logic-Algorithms, Techniques and Implementations*, 1st ed.; InTech Ltd.: Croatia, 2012.
13. Fanelli, A. M.; Pedrycz, W.; Petrosino, A. *Fuzzy Logic and Applications*, 1st ed.; Springer: Verlag Berlin Heidelberg, 2011.
14. Jou, J. M.; Chen, P-Y.; Yang, S-F. An Adaptive Fuzzy Logic Controller: Its VLSI Architecture and Applications. *IEEE Trans. Very Large Scale Integr. (VLSI) Syst.* **2000,** *8*, 52. ISSN: 1063-8210.
15. Huiwen, Y. W.; Chen, D. Z. Adaptive Fuzzy Logic Controller with Rule-Based Changeable Universe of Discourse for a Nonlinear MIMO System. In *Proceedings of International Conference on Intelligent Systems Design and Applications*. ISBN: 0-7695-2286-6, 2005.
16. Yas, Q. M.; Khalaf, M. Reactive Routing Algorithm Based Trustworthy with Less Hop Counts for Mobile Ad-Hoc Networks Using Fuzzy Logic System. *J. Southwest Jiaotong Univ.* **2019,** *54* (3), 1–11.
17. Inaba, T.; Sakamoto, S.; Kulla, E.; Caballe, S. An Integrated System for Wireless Cellular and Ad-Hoc Networks Using Fuzzy Logic. In *2014 International Conference on Intelligent Networking and Collaborative Systems IEEE*, 2014; pp 157–162.
18. Balan, E. V.; Priyan, M. K.; Gokulnath, C.; Devi, G. U. Fuzzy Based Intrusion Detection Systems in MANET. *Procedia Comput. Sci.* **2015,** *50*, 109–114.

19. Yas, Q. M.; Zadain, A. A.; Zaidan, B. B.; Lakulu, M. B.; Rahmatullah, B. Towards on Develop a Framework for the Evaluation and Benchmarking of Skin Detectors Based on Artificial Intelligent Models Using Multi-Criteria Decision-Making Techniques. *Int. J. Pattern Recogn. Artif. Intell.* **2017,** *31* (03).

20. Yas, Q. M.; Zaidan, A. A.; Zaidan, B. B.; Rahmatullah, B.; Karim, H. A. Comprehensive Insights Into Evaluation and Benchmarking of Real-Time Skin Detectors. *Review, Open Issues & Challenges, and Recommended Solutions Measurement* **2017,** *114,* 243–260.

21. Chaudhary, A.; Kumar, A.; Tiwari, V. N. A Reliable Solution against Packet Dropping Attack Due to Malicious Nodes Using Fuzzy Logic in MANETs. In *2014 International Conference on Reliability Optimization and Information Technology (ICROIT)* IEEE, 2014; pp 178–181.

22. Anuradha, M.; Mala, G. S. A. Cross-Layer Based Congestion Detection and Routing Protocol Using Fuzzy Logic for MANET. *Wireless Netw.* **2016,** *23* (5), 1373–1385.

23. Robinson, Y. H.; Rajaram, M.; Julie, E. G.; Balaji, S. Detection of Black Holes in MANET Using Collaborative Watchdog with Fuzzy Logic. *World Acad. Sci. Eng. Technol. Int. J. Comput. Electr. Autom. Control Inf. Eng.* **2016,** *10* (3), 622–628.

24. Kennedy, J.; Eberhart, R. Particle Swarm Optimization. In *Proceedings of the IEEE International Joint Conference on Neural Networks,* 1995; pp 1942–1948.

25. Li, X.; Yao, X. Cooperatively Coevolving Particle Swarms for Large Scale Optimization. *IEEE Trans. Evol. Comput.* Apr **2012,** *16* (2).

26. Chandra Mohan, B.; Baskaran, R. Survey on Recent Research and Implementation of Ant Colony Optimization in Various Engineering Applications. *Int. J. Comput. Intell. Syst. Taylor and Franchis* **2011,** *4* (4).

27. Zhang, Z.; Long, K.; Wang, J.; Dressler, F. On Swarm Intelligence Inspired Self-Organized Networking: Its Bionic Mechanisms, Designing Principles and Optimization Approaches. *Commun. Surv. Tutor.*, IEEE Jul 4, **2013,** (99).

28. Akay, B.; Karaboga, D. *A Modified Artificial Bee Colony Algorithm for Real-Parameter Optimization,* Vol. 192; Elsevier, 2010.

29. Fana, S-K. S.; Changa, J-M. A Parallel Particle Swarm Optimization Algorithm for Multi-Objective Optimization Problems. *Eng. Optim. Taylor and Franchis* **2009,** *41* (7).

30. Tonguz, O. Biologically Inspired Solutions to Fundamental Transportation Problems. *IEEE Commun. Mag* Nov **2011,** *49* (11), 106–115.

31. Yang, J.; He, L. F.; Fu, S. Y. An Improved PSO-Based Charging Strategy of Electric Vehicles in Electrical Distribution Grid. *Appl. Energy* **2014,** *128,* 82–92.

32. Zhang, Y.; Wang, S.; Ji, G. A Comprehensive Survey on Particle Swarm Optimization Algorithm and Its Applications. *Hindawi Pub. Corp. Math. Probl. Eng.* Feb **2015,** *2015,* Article ID 931256.

CHAPTER 11

Enhanced Intelligence Architecture

AMBIKA N.*

Department of Computer Science and Applications, St. Francis College, Bangalore, India

E-mail: ambika.nagaraj76@gmail.com

ABSTRACT

Many smart sensors and actuators come together to make an Internet of Things (IoT) network. Artificial intelligence (AI) is a methodology that enables machines to analyze the received data. Based on the outcome, they instruct the devices to follow a set of instructions. Blockchain is the technology that brings trust to the person dealing with it. These different domains are combined with storage to enable a combo set of features. The previous methodology provides a reliable, secure system. The user has to feed his username and password to use the IoT devices. The data are transmitted to its peer devices using storage. A calculated outline of blockchain and artificial intelligence for IoT is where the IoT stage depicts a mix of six layers—physical, correspondence tier, connect control, administration, executives, and application tier. The physical distinguishes the information or data like temperature, area, contamination, climate, movement, and farming. The data are assembly from different sensor gadgets like RFID, barcode, and infrared. This region has various safety dangers and problems, for example, moving the data starting with one spot then onto the next. It makes it unstable for malevolent people. Blockchain and AI usage make the data or exchange of bitcoin. The gathered information moves to

Computational Intelligence Applications for Software Engineering Problems. Parma Nand, PhD, Rakesh Nitin, PhD, Arun Prakash Agrawal, PhD & Vishal Jain, PhD (Eds.)
© 2023 Apple Academic Press, Inc. Co-published with CRC Press (Taylor & Francis)

the correspondence layer. It is a model for moving data with one device then onto the next gadget. It finishes by actualizing a few cutting-edge innovations, for example, WiFi, Zigbee, radio, and infrared movement. This region has safety and protection problems, blockchain and AI innovation use in point-to-point organizations, and omnipresent broadband use for encoding and verification. The suggestion adds some of the features to the previous system and increases security by 7.37%, reliability by 1.83%, and speed by 8.053%.

11.1   INTRODUCTION

Many smart sensors[27] and actuators come together to make an Internet of Things (IoT) network.[9] These elements communicate with each other using a common platform. IoT (Nagaraj 2021) is not one technology. It is a thought during which several new things have gotten networked and attached anytime, anywhere, with something, and anyone estimation mistreatment any route or communication system and any assistance during a sophisticated atmosphere. It is a vibrant international interface foundation with self-configuring capacities. It is supported by practical conversation protocols wherever physical and realistic belongings have personalities, physical property, and virtual attributes and use reasoning intersection and area units consistently blended into the knowledge system.

Storage computing[2,8] is the prototype for facultative available. The on-demand system accesses the distributed store. It is free with negligible administration energy or storage supplier intercommunication. Artificial intelligence[7,19] is a methodology that enables machines to analyze the received data. Based on the outcome, they instruct the devices to follow a set of instructions. Blockchain[20] is the technology that brings trust to the person dealing with it. These different domains are combined to enable a combo set of features.

The previous work[35] uses blockchain.[4,10] The methodology provides reliable, secure system. The user has to feed his username and password to use the IoT devices. The data are transmitted to its peer devices using storage. A calculated outline of blockchain and AI for IoT is where the IoT stage depicts a mix of six layers—physical, correspondence tier, connect control, administration, executives, and application tier. The physical distinguishes the information or data like temperature, area, contamination, climate, movement, and farming. The data are assembly from different sensor gadgets like RFID,[13,16] barcode, and infrared.[26] This region has various safety dangers and problems, for example, moving the data starting with one spot then onto the next. It makes it unstable for malevolent people. Blockchain and AI usage make the data or exchange of bitcoin. The gathered information moves to the correspondence layer. It is a model for moving data with one device then onto the next gadget. It finishes by actualizing a few cutting-edge innovations, for example, WiFi,[12,34] Zigbee,[17,30] radio, and infrared wave. This layer has security and protection issues, blockchain and AI innovation use in point-to-point organizations, and omnipresent broadband use for encoding and verification. The suggestion adds some of the features to the previous system and increases security[1] by 7.37%, reliability[3] by 1.83%, and speed[25] by 8.053%.

The chapter is divided into five divisions. The literature survey details the summary of the work suggested by various authors in the second segment. The proposed work is detailed in section three. The work is analyzed in division four. The work concludes in the fifth segment.

11.2 LITERATURE SURVEY

The previous contribution[35] uses the three domain. The work includes four-tier—the storage, blockchain, AI bring decentralization, and security to the network. Fog communicates using AI and blockchain methodology. It uploads the data into the fog layer. Edges installed at the server end are AI-enabled. They transfer the data to the edge devices. IoT devices use device intelligence to store the data at the device layer. The proposal is enhancing the previous work. The methodology provides reliable, secure system. The user has to feed his username and password to use the IoT devices. The data are transmitted to its peer devices using storage. A calculated outline of blockchain and AI for IoT is where the IoT stage depicts a mix of six layers—physical, correspondence tier, connect control, administration, executives, and application tier. The physical distinguishes the information or data-like temperature, area, contamination, climate, movement, and farming. The data are assembly from different sensor gadgets like RFID, barcode, and infrared. This region has various safety dangers and problems, for example, moving the data starting with one spot then onto the next. It makes it unstable for malevolent people. Blockchain and

AI usage make the data or exchange of bitcoin. The gathered information moves to the correspondence layer. It is a model for moving data with one device then onto the next gadget. It finishes by actualizing a few cutting-edge innovations, for example, WiFi, Zigbee, radio, and infrared wave. This layer has security and protection issues, blockchain and AI innovation use in point-to-point organizations, and omnipresent broadband use for encoding and verification.

The service design[38] of the IoT-storage system has trio regions. They are the assortment, process, and application service region. The sting network lies within the information assortment region. The sting platform is within the process region. The program assistance region and the storage are part of the appliance assistance region. Their measure has three service exhibitions within which the user needs square measure restricted to IoT. Several mixed-service needs square measure supported. The multiple IoTs express as mixtures of primary scenes. The good surroundings applications like good transportation, healthcare, and increased reality are examples. These good surroundings practices have several users and receive an oversized variety of recurrent/same needs. The measure suggests supervising boundary devices: the unmoving condition and mobile status. The border devices measure is located in the fastened positions of its height. The edge nodes have a set moving range in the case of moving nodes. It desires fewer edge nodes and has a slightly weaker period performance. Internal attacks do not happen the entire instance. The moving mode may be an acceptable selection of inbound things.

The planned[11] combined, ensured, and design for the IoT and, storage calculation envisages to supply assured good employment and utilization anyplace, anytime, any steady, any pattern, and any system freelance of any implicit application with one IoT-enabled clever revolving credit. The clever arrangement means information at the entryway that transfers the required collection to the storage using IP/MPLS center system. The data on the ISC are end-to-end the planet at totally various information centers. It reinforces the validation method and accessibility of knowledge anytime. The technique embraces elliptical curvature cryptology and computer code-based digital certificates. It makes validation, private, secrecy, and sincerity. The intended smart revolving credit is an associate degree IoT-enabled operational record that adheres to the ISO/IEC customary. ISC fashioned for the planned design has an associate degree RFID tag, biometric model icon, and distinctive figure because of some options.

ISC takes the scheme on correspondence method to make safety and use employment exploitation one UID per subject. The planned design adopts two forms of readers. RFID reader mold to the ISO customary and acts as an associate degree interface between ISC and entry. The planned entry in the heart of the setting has a direct approach to the knowledge center to recover the consumer info as shortly because it finds the UID from the ISC through the reader. This planned good entry is incredibly versatile and configurable to affirm different application needs and supports many rules. The planned design performs authentication methods at four levels. Mutual authentication happens between the ISC and reader.

The projected IoT-BSFCAN[15] forms associate degree intelligent design for property IoT employing a mechanics of the gearing-develop system. This system is a fundamental suggestion for 5G-enabled movable-fog computation assistance that is thought of as a nonpublic IoT-storage scheme instead of ancient fog computation. This amendment provides less livelihood prices to realize higher adaptability with payload reconciliation. The gear–train network uses a location site to put the network property in a storage-associated fog server to cut back the deployment value of a brand new nonpublic IoT-storage system. It is capable to supply many customer link that uses sensible mechanistic fogging to travel as a fog-host. It permits the electronic equipment to keep up the continual connection while not the intake of excess-depot areas. It is one in every one of the credential options of sensible movable fogging. Moreover, the information ware-house is often dispersed across the various gear–train networks to traverse the terminal access. A query-driven model is developed as an immediate mobile-fogging to transfer the payload between the house entrances. Since moveable-fogging deals with the communication of knowledge admittance over open/secret IoT storage, it is going to expertise a lot of server admittance to promote system access. The projected IoT-BSFCAN provides circumstance-delicate watching scheme to confirm the credential goals. The detector tier accumulates sensitive information from the period atmosphere and communicates the accumulated knowledge to the fog computing layer.

It is a unique cooperative and sensible network based invasion detection system design[29] to effectively defend against system-related internet assault in SDN-based storage device networks. This safety design comprises of a graded arrangement of NIDS point, Edge-IDS, Fog-IDS, and storage-IDS. These IDS measures supported the automobile education/deep education formula. The people are set within the same computing layer that is a distributed style. The trendy storage-based IoT systems square measure is usually classified into three significant levels: edge computation, fog calculation, and storage computation that correspond to a few organize of IoT scheme—conceptualization, spatial arrangement system, and utilization.

The work[14] achieves a sensible meter with a laterally automatic machine identification performance. It consists of a hardware tier, a knowledge method tier, and a recognition tier. The home and cognizant grid domains square measure lined by the appliance obtained from this measure arrangement. The hardware layer is a sensible meter. It is responsible for electronic energy signals, wave shape correction, and regulation. It provides the info method tier with primary voltage and current signals. The info method tier is the advanced processing half within the STM32. It includes internal wave shape extraction, noise reduction, and state detection. It is responsible for providing an electronic appliance characteristic process for the electronic appliance recognition rule. The popularity tier includes a wave shape recognition rule, info creation, and segmentation. It is responsible for scheming and distributing the electronic power components of the data method layer. It uses fewer characteristic teams for the gradable organization and separates different characteristics of the info matrix. It provides improved confirmation info for the core wave shape. It has an affidavit rule to scale the time complexness of identification estimation.

The system design[36] consists of five layers. The resource tier provides every kind of resource concerned within the total existence series of production and primary reservoir of the producing assistance program. It also has hardware-producing resources, procedure sources, and alternative resources. Perception tier realizes intelligent perception and identification of producing reservoir through various detection instruments and changes within the production existence interval, thereby providing sturdy assistance for activity platforms to showing intelligence to establish and manage to manufacture sources. Detection instruments embrace deuce-magnitude barcode, RFID interpret, sensing element, video seizure, and GPS. The arranger considers a package system convergence arranger, detector arrangers, framework arrangers, information arrangers, system arrangers, retention arrangers, technological asset arrangers, and different arrangers. System tier supply all required bearer networks for admittance to numerous reservoirs in merchandise whole life cycle, together with

2G systems, 3G systems, 4G systems, satellite systems, cable systems, company internal wireless systems, etc. The service tier principally gives two classes of assistance. The CMfg assistance is a structure functional assistance. The previous is the consequence of assistance encapsulation of producing reservoir and capability. Utilization tier refers to the on-demand use of varied Mfg assistance within the complete existence interval, together with style, commercial enterprise, research, modeling, administration, livelihood, employment, etc.

The work[39] proposes a practical design for a high-level dispatch site framework. It has five sections—the canny specialized region, the shrewd dispatching district, the astute flying and platform region, the wise order and dominance framework, and the keen investigation appraisal framework. The five sections comprise the foundation, offices, gear, equipment, and programming. It incorporates the entire mission cycle of the ground and dispatch frameworks from flight articles passage to dispatch. The compositional structure has insightful elements of the parts. The system characterizes as the interrelation and the convergence of the components. It includes dispatch conveyance and flying loading. It incorporates tetrad tiers: the concrete tier, the discernment tier, the organization tier, and the utilization tier. The actual tier incorporates the items and actuators of the dispatch site. The discernment tier comprises the sensing element and information handling framework. The organization tier supplies the entrance passages and spine organization. The utilization tier serves utilization frameworks through the middleware stage. The center of the wise framework is the regulator of the programmed activity framework crossing the tetrad tiers. The examination assembles IoT structure, storage stage, middleware, admittance passage, and programmed dominance framework.

The proposed is multilayer engineering[18] engaged by collaborative machine learning. It conveys constant wellbeing observing and irregularity identification. The ECG communication bears the registered information from 12 advantages, which are acquired utilizing terminals connected to the outside of the organic structure. Each lead gauges a particular electric potency, a perspective on the heart's electrical activity in various areas in the organic structure. The presentation of ANN has a support vector machine classifier. This assessment causes us to distinguish the effectiveness of each framework. The yield of classifiers affirmed that the ANN framework has higher exactness, quality, and affectability. The SVM calculation is moderately troublesome to execute in low-end gadgets and is more procedure escalated. The subsequent methodology has two gatherings of information—ordinary and irregular beats.

The center part that deals with all connected storage-based creation administrations[31] is an organized business measure organization stage. This stage recognizes, forms, and executes creation benefits are given by various working together gatherings inside a worth organization. This stage is an organized framework by the central association inside the production network. For the situation that the inventory network designs of a worth organization incorporate extraordinary associations, the organized business measure the board will utilize administration geography advancements rather than coordination innovations. IoT-empowered administrations can be classified into four classes. Web of assembling administrations addresses all assets and capacities virtualizes and epitomized inside administrations, such as gear, shrewd workshops, 3D printers, and savvy industrial facilities. Web of coordination administrations incorporates adjusted transportation and stock administration administrations like keen trucks, transportation specialist organizations empowered by ongoing area frameworks, and keen stockrooms. It forms the web of up-transfer coordination administrations. It is dependent on an item or administration request conjecture by a central association. The piece of these administrations is taken care of by configuration time coordination plans. The Internet of down-transfer coordination administrations forms clients' genuine interest. Web of clients depicts that clients are empowered to make esteem dependent on proposed items and administrations inside a worth organization. It acknowledges by cocreation instruments that emphasize overseeing clients experience the board by detecting, investigating, and following up on their passionate, cognitional, and social data.

The architecture[6] plans for the system. It consists of four tiers ranging from very cheap with RFID/detector system. The access entrance is completely different access networks like the net. The middleware has divergent services like catalog assistance, connection modeling and administration, and content supervision. Connection splits into components context suppliers like sensors, context client or context-aware services, and context broker. These components are superimposed to a circumstance supplying structure, wherever the interaction among these components follows a fixed pattern, wherever the circumstance client asks for data from a circumstance supplier through circumstance agent.

The circumstance agent seeks circumstance data from different services on the storage. The circumstance data have varied levels of caliber of circumstance and caliber of assistance. QoC refers to data and not the method nor the instrumentation part. The circumstance data are mass from the sensing element diffused within the surroundings and attached to the storage/SaaS. It specifies SaaS above all since it is the most appropriate. Their characteristic properties gather similarly. The circumstance data, attended with quality data, are passed on to the connected instrument. The action handler attempts to communicate with the user and IoT good surroundings, choosing from completely various action style. The condition is entire mechanistic, chosen once an advanced degree of QoC and QoS. The improvement approach organization is shrewd surroundings that can counsel action to take per the connection data provided wherever the connected degree of QoS or QoC does not seem to be ideal. Finally, contact or data displays on the computer program to express communication with a decreased level.

IoT instrumentation representative[24] features a deposit to stock IoT instrumentation info from store-scale IoT assistance. Most of the present-day, IoT services square measure supply arthropod genus to admittance their devices. The IoT instrumentation representative detains the customer's document to admit those storage services through algorithms like OAuth. It preserves instrumentation info on the deposit. Once saving device info from storage-scale IoT services, IoT instrumentation representatives made information transmission to modify province values from every instrument. The information transmission is going to be accustomed initiate values in purpose. The information transmission is an associate degree abstraction entity wherever information from the device square measure streamed. Transmission could have a completely different data supply concerning the services that the user has signed. Or else, the supply of fundamental measure information transmission is also storage service like weather.com. To create an interface the information transmission delineates concerning a service, not a tool. With the information transmission, the user will access every service directly while not regarding device capabilities. IoT instrumentation representative provides service abstraction arthropod genus for services. This arthropod genus square measure is known as IoT instrumentation harmony API. IDH API outlines as REST, and the construction of IDH can be acquired as associate degree XML representation. Every customer and instrument in IoT Instrumentation represents with UUID.

To create an IoT-storage foundation, the reproduction of detected knowledge is vital. The detected data repetition makes a positive response for the actuators in proceeding of any host-aspect failure. The backing scheme[21] will take over the management. For storage, the device of available geographical area for the substitute employs to handle any disaster or failure within the storage server. The projected multilevel design is reliable. It does not have servers on a similar level with detected knowledge replication. The quantity of knowledge for a similar application might vary on mist, fog, and storage levels. However, knowledge would exist on all the amount required over the application management. The portable agent[23,32] works as a resource and system watching broker. It distributes the applying and tie-state data with different factor on various hosts within the organization. They are accountable for assigning priority to the IoT applications relying upon their delay tolerance. Within compact failure or compact scheduled conclusion, these priority data employ at the time of burden arrangement of a selected server. It additionally helps in new path discovery when the load distribution. It checks on rhythmical observation and substitute of knowledge.

The electrocardiogram detecting system[40] is the basis of the whole scheme. It is answerable for assembling functional facts from the body exterior and transmittal this information to the IoT storage through wireless communication. The electrocardiogram signals measure processed data using a sequence of processes, like augmentation, clarifying, etc. It improves the indication value and fulfills the necessities of wireless transmission. The electrocardiogram information collected from devices square measure communicated to the IoT storage via a particular wireless procedure, for example, WiFi, Bluetooth, Zigbee, etc. All conventions will give enough information rates for transmittal electrocardiogram signs with sustaining vitality depletion. A wise terminal is sometimes required to receive the electrocardiogram information then send the information to the IoT storage through the wireless procedures of the overall container receiver provision[5,22] or future evolution (LTE). IoT storage for electrocardiogram observation sometimes consists of four purposeful modules, that is, information improvement, information storage, information inquiry, and malady warning. The vital options extract from electrocardiogram signals discover potential heart diseases. The graphical user interface is answerable for information visual image and administration. It provides easy accessibility to the information within the IoT storage. Users will enter onto the storage to amass envisioned electrocardiogram information in real time.

The planned system[37] has vital potential for analyzing patients' attention knowledge to avoid decease environments. This structure assembles the long-suffering knowledge exploitation of numerous pharmaceutical appliances and sensors. To prevent fraud or clinical mistakes by wellbeing authorities, methods like gesture sweetening and watermarking are part of this skeleton. During this network, the patient will monitor employing an assortment of tiny-powdered and lightweight device networks. The scheme supplies an associate in the nursing boundary between the surgeon and the long-sufferings for dual announcement. Associate in the nursing design of u-healthcare contains body space network, sensible pharmaceutical server, and clinic arrangement area unit taken as a necessary element to outline this technique. The abstract structure of IoT is a sensible student interactive attention arrangement that consists of three stages. In stage 1, learners' health knowledge was nonheritable from healing machines and sensors. The nonheritable learning relays to a storage scheme employing an entranceway or native process unit. In stage 2, the remedial dimensions area unit employed by opinion structure to require psychological feature call associated with scholar wellness. Stage 3 provides warnings to caretakers.

The device layer[33] watches the varied open organization situations. It directs the cleaned evidence spent domestically to the fog level and customs the demand facilities. The device-level communicates the cleaned facts to the fog level. It contains of a superior scattered SDN organizer. Every fog device conceals the little-related communal and is chargeable for information investigation and repair conveyance promptly. The fog level gives the consequences of administered production information to the storage and device layers if required. The fog layer provides localization, whereas the storage layer provides widespread-zone observation and management. It is huge-gauge happening recognition, behavioral investigation, and extended arrangement acknowledgment by providing dispersed figuring and storing. It is a disseminated storage supported by the blockchain practice that gives protected, low-charge, and on-request entree to the foremost reasonable figuring organizations. The purchasers will explore, discover, deliver, usage and mechanically unencumbered all the calculating possessions, like servers, data, and applications, they need. In the fog node, all the SDN controllers connect in a distributed manner. Each SDN controller scepters by the associate investigation function of the course decree and a container movement. It secures the network throughout inundation assaults. The authors deployed multi-interfaced sink nodes fortified with associate SDN switches. It facilitates the new wireless announcement technologies supported by IoT. To mixture all of the facts streams originating from native IoT instruments, we tend to thought-about the multi-interfaced BSs to represent a wireless entranceway. It acts as forwarding arranges for the SDN supervisors of the fog device. It screens movement at the information plane and launches worker assemblies. The SDN controllers within the fog device additionally give software design boundaries to system organiza-tion workers to supply numerous vital interacting abilities. A fog device will admit the disseminated storage over the web to amenably organize the appliance service and computing availableness. Fog devices will offload their computing assignments to the scattered storage once they do not have decent figuring possessions to method their native information streams at the expense of enlarged potential in infrastructures and system reserve depletion.

11.3   PROPOSED WORK

In the previous work,[35] blockchain was used. The methodology provides reliable, secure system. The user has to feed his username and password to use the IoT devices. The data are transmitted to its peer devices using storage. A calculated outline of blockchain and AI for IoT is where the IoT stage depicts a mix of six layers—physical, correspondence tier, connect control, administration, executives, and application tier. The physical distinguishes the information or data like temperature, area, contamina-tion, climate, movement, and farming. The data are assembly from different sensor gadgets like RFID, barcode, and infrared. This layer has various security dangers and issues, for example, moving the data starting with one spot then onto the next. It makes it unstable for malevolent people. Blockchain and AI usage make the data or exchange of bitcoin. The gathered information moves to the correspondence layer. It is a model for moving data with one device then onto the next gadget. It finishes by actualizing a few cutting-edge innovations, for example, WiFi, Zigbee, radio, and infrared wave. This level has safety and protection problems, blockchain and AI innovation use in point-to-point organizations, and omnipresent broadband use for encryption and verification.

- Analytics intelligence—an enormous measure of information in IoT solicitations is created or gathered by billions of detecting gadgets in different arenas. These gadgets give the yield as critical information streams. Investigation insight to these information streams discovers a new data environment. It provides a future situation in IoT submissions. Control choice is a fundamental technique for IoT information stream. It gives the data quality. Subsequently, AI is useful for IoT claims like the keen city, brilliant automobile, and shrewd medical care, and it is a piece of profound education. The assembly of blockchain and AI for IoT addresses an excellent chance for remote and communal areas. The area can do misusing these advancements. It gets an opportunity to improve the current cycle. It finds new business procedures for creating imaginative administrations in another age of shoppers. It depends on the creation of novel contemplations or data. Examination knowledge and blockchain innovation changes over conditions like emergency clinics, air terminals, schools. The dynamic capacities and learning abilities are utilized for all activities rapidly and safely.
  The current proposal has an intrusion detection layer in the storage. This layer has the set of legitimate behavior of IoT devices. When the device communicates with the storage, its behavior is analyzed making a comparison with the behavior set stored on them.
- Digital identity—it is an arising technique utilized for different IoT applications. It mostly gives a one-of-a-kind distinguishing proof location to the gadgets. All gadgets cooperate progressively, secure, and decentralized way. An organization such as Flipkart and e-straight causes us to detect the parcel from consignment to conveyance. The scanner tag on the bundle is the computerized delivering character. It is the vital impression for instruments, automatic mechanism framework, and shrewd vehicles. The grouping of blockchain and AI for IoT gives a computerized encoding of the knowledge or converts the information into an advanced structure since it is an encryption part. The idea utilized

in the administration and business layer gives the executives of information, data, and applications. Using AI and blockchain for IoT, miniature robots are applied to pesticides and herbicides. It catches the detailed execution of each plant. It gives financial, working points of interest like enhanced steadiness in the yield, declines produces due to infection, the period and exertion essential and decreases the consumption and misuse of insecticides and herbicides.

The suggestion includes generating a hash code (using blockchain concept to improve trust) using location and device identification. Using these parameters in deriving hash codes increases reliability of the systems.

- Disseminated storage—it is a centerpiece of blockchain innovation, and it puts away, the data are circulated and decentralized structure in blockchain networks. This approach gives overall discernibility, truthfulness to the storage in a fake astute way. Conveyed distributed storage offers added reply for an information base question that improving the scope of evidence from IoT applications by the union of blockchain and AI in IoT. The expectation idea in AI has a pivotal part (forecast) disseminated learning ahead of period and mentions using the evidence later on. Dispersed distributed storage utilized in the assistance layer offers assistance like a dataset, dynamic to the IoT applications. The savvy contract exchanges in shared arrangements without the utilization of a unified center point. This cycle additionally gives a protected, faster, reliable method of correspondence for one another.

The suggestion that has a centralized data system is managed to contain the hash code of the transaction and where it is available. Using, this methodology increases speed in communication. The users are also given the flexibility to use their devices from multiple locations.

- Decentralization—IoT gadgets are employing public organizations. It is compromised by any third pernicious client because IoT structures employ focused employee philosophy. The use of blockchain settles the safekeeping matter in IoT surroundings. It makes conventional documents that are itemized consistently and are distributed. The union of blockchain and AI for IoT apply many arrangement resolutions like evidence of work, confirmation of stake. IoT submissions have a preliminary job in educating the collection of evidence from diverse devices to make the AI structure. In this cycle, safekeeping, shield, and energy employment are critical complications, with the assembly of blockchain and AI take care of these topics.

- Authentication—confirmation is any technique that confirms a portion of the safe conversation. It is the clue employed habitually in IoT submissions like scientific, brilliant automobiles, and others. The IoT claim uses a concentrated way in this interaction, and it is matter to the panel. Blockchain invention gives the finest attitude to automated discussion with one individual. The cryptographically noticeable and confirmed by all diggers. It consumes digital coinage, for example, bitcoin. Simulated intelligence uses dynamic capacities for a particular transaction. The combination of blockchain and AI for IoT gives the design known as regionalized AI. It is employed for an automated interchange in a safe, valid by excavators. Validation and checked are in the communication layer. With the use of blockchain innovation in IoT, devices and passage may secure data. All data are confirmed in the disseminated hyper record. The hubs' move and can agree to the dependability before bearing them.

- Chain structure—it is a collection of evidence in IoT use. It is produced by diverse perceiving appliances, for example, mobile phone, WiFi, and pen drives. The genuine level or insight level has a sequence organization identifies with the dataset employing shrewd settlement, hash measurements, wide-reaching enrolment, and carried character for blockchain frame capability of IoT chain organization. Merkle tree dispersion of squares in blockchain innovation. The discernment layer distinguishes the information/data like temperature, area, contamination, climate, movement, agribusiness, and others from sensor gadgets. This layer has various types of security dangers. The work uses the idea of blockchain and simulated intelligence (Table 11.1).

TABLE 11.1   List of Enhanced Features in the Present Proposal.

Component	Addition feature
Analytics intelligence	It has an intrusion detection layer in the storage. This layer has the set of legitimate behavior of IoT devices. When the device communicates with the storage, its behavior is analyzed making a comparison with the behavior set stored on them
Digital identity	It includes generating a hash code (using blockchain concept to improve trust) using location and device identification. Using these parameters is deriving hash codes increases reliability of the systems
Distributed storage	The suggestion has a centralized data system is managed containing the hash code of the transaction and where it is available

11.4   ANALYSIS OF THE WORK

11.4.1   INTRODUCING INTRUSION DETECTION LAYER

Anomaly detection method enhances security in the network. A dataset of behaviors is maintained and analyzed against the behavior evaluated by the storage at every instance. Comparing with the previous study, the suggestion increases security by 7.37%. The same is represented in Figure 11.1.

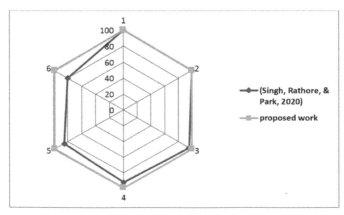

FIGURE 11.1   Comparison of security in the network.

11.4.2   ENHANCING TRUST IN THE NETWORK

Reliability is essential parameter in the network. Both the communicating parties require trust among themselves to share confidential information. It includes generating a hash code (using blockchain concept to improve trust) using location and device identification. The suggestion aims to improve trust among the transmitting entities by 1.83% compared to previous work.[35] The same is represented in Figure 11.2.

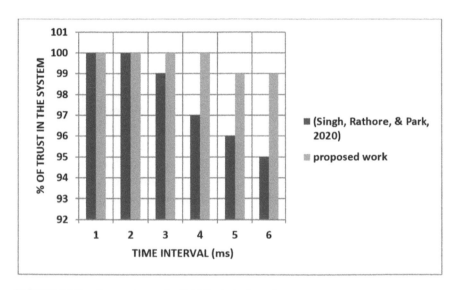

FIGURE 11.2   Comparison of reliability in both systems.

11.4.3   INCREASING SPEED

The system with good speed is always under consideration. The suggestion that has a centralized data system is managed by containing the hash code of the transaction and where it is available. This methodology increases speed in communication. The users are also given the flexibility to use their devices from multiple locations. The speed to the suggestion is increased by 8.053% compared to the previous work.[35]

11.5   CONCLUSION

The work designs the architecture where IoT devices, storage, edge systems, and fog devices communicate to accomplish the job. The previous work uses blockchain. The methodology provides reliable, secure system. The user has to feed his username and password to use the IoT devices. The data are transmitted to its peer devices using storage. A calculated outline of blockchain and AI for IoT is where the IoT stage depicts a mix of six layers—physical, correspondence tier, connect control, administration, executives, and application tier. The physical distinguishes the information or data like temperature, area, contamination, climate, movement, and farming. The data are assembly from different sensor gadgets like RFID, barcode, and infrared. This layer has various security dangers and issues, for example, moving the data starting with one spot then onto the next. It makes it unstable for malevolent people. Blockchain and AI usage make the data or exchange of bitcoin. The gathered information moves to the correspondence layer. It is a model for moving data with one device then onto the next gadget. It finishes by actualizing a few cutting-edge innovations, for example, WiFi, Zigbee, radio, and infrared wave. This layer has security and protection issues, blockchain and AI innovation use in point-to-point organizations, and omnipresent broadband use for encryption and verification. The suggestion adds some of the features to the previous system and increases security by 7.37%, reliability by 1.83%, and speed by 8.053%.

KEYWORDS

- **IoT**
- **artificial intelligence**
- **blockchain**
- **storage computing**
- **architecture**
- **reliability**
- **security**
- **availability**

REFERENCES

1. Abusukhon, A.; M. Talib. A Novel Network Security Algorithm Based on Private Key Encryption. In *International Conference on Cyber Security, Cyber Warfare and Digital Forensic (CyberSec)*; IEEE: Kuala Lumpur, Malaysia, 2012; pp 33–37.
2. Aceto, G.; de Donato, A. B.; Pescape, A. W. Cloud Monitoring: A Survey. *Comput. Netw.* **2013**, 2093–2115.
3. Ahmad, M. Reliability Models for the Internet of Things: A Paradigm Shift. In *IEEE International Symposium on Software Reliability Engineering Workshops*; IEEE: Naples, Italy 2014.
4. Al Omar, A.; Rahman, M. S.; Basu, A.; Kiyomoto, S. Medibchain: A Blockchain Based Privacy Preserving Platform for Healthcare Data. In *International Conference on Security, Privacy and Anonymity in Computation, Communication and Storage*; Springer: Cham, Guangzhou, China, 2017; pp 534–543.
5. Al-Ali, A. R.; Zualkernan, I.; Aloul, F. A Mobile GPRS-Sensors Array for Air Pollution Monitoring. *IEEE Sens. J.* Oct **2010**, *10* (10), 1666–1672.
6. Alhakbani, N.; Hassan, M. M.; Hossain, M. A.; Alnuem, M. A Framework of Adaptive Interaction Support in Cloud-Based Internet of Things (IoT) Environment. In *International Conference on Internet and Distributed Computing Systems*; Calabria, Italy, 2014; pp 136–146.
7. Allam, Z.; Dhunny, Z. A.. On Big Data, Artificial Intelligence and Smart Cities. *Cities* **2019**, *89*, 80–91.
8. Ambika, N. Encryption of Data in Cloud-Based Industrial IoT Devices. In *IoT: Security and Privacy Paradigm*; Pal, S., Díaz, V. G., Le, D-N., Eds.; CRC Press, Taylor & Francis Group: UK, 2020; pp 111–129.
9. Ambika, N. Energy-Perceptive Authentication in Virtual Private Networks Using GPS Data. In *Security, Privacy and Trust in the IoT Environment*; Mahmood, Z., Ed.; Springer: Cham, Switzerland, 2019; pp 25–38.
10. Atlam, H. F.; Wills, G. B. *Technical Aspects of Blockchain and IoT.*, Vol. 115. In *Role of Blockchain Technology in IoT Applications*, 2019.
11. Bai, T. D. P.; Rabara, S. A. Design and Development of Integrated, Secured and Intelligent Architecture for Internet of Things and Cloud Computing. In *3rd International Conference on Future Internet of Things and Cloud*; IEEE: Rome, Italy, 2015; pp 817–822.
12. Balasubramanian, A.; Mahajan, R.; Venkataramani, A. Augmenting Mobile 3G Using WiFi. In *Proceedings of the 8th International Conference on Mobile Systems, Applications, and Services*; ACM, 2010; pp 209–222.
13. Bhuptani, M.; Moradpour, S. *RFID Field Guide: Deploying Radio Frequency Identification Systems*; Prentice Hall PTR, 2005.
14. Chen, S. Y.; Lai, C. F.; Huang, Y. M.; Jeng, Y. L. Intelligent Home-Appliance Recognition over IoT Cloud Network. In *9th International Wireless Communications and Mobile Computing Conference (IWCMC)*; IEEE: Sardinia, Italy, 2013; pp 639–643.

15. Deebak, B. D.; Al-Turjman, F.; Aloqaily, M.; Alfandi, O. IoT-BSFCAN: A Smart Context-Aware System in IoT-Cloud Using Mobile-Fogging. *Future Gen. Comput. Syst.* 2020, *109*2020, 368–381.

16. Dimitriou, T. A Lightweight RFID Protocol to Protect aganist Traceability and Cloning Attack. In *1st International Conference on Security and Privacy for Emerging Areas in Communications Networks*; IEEE, 2005.

17. Faludi, R. *Building Wireless Sensor Networks: With ZigBee, XBee, Arduino, and Processing*; O'Reilly Media, Inc, 2010.

18. Farahani, B.; Barzegari, M.; Aliee, F. S.; Shaik, K A. Towards Collaborative Intelligent IoT eHealth: From Device to Fog, and Cloud. *Microprocessors and Microsystems* 2020, *72*, 1–16.

19. Farivar, F.; Haghighi, M. S.; Jolfaei, A.; Alazab. M.; Artificial Intelligence for Detection, Estimation, and Compensation of Malicious Attacks in Nonlinear Cyber-Physical Systems and Industrial IoT. *IEEE Trans. Indust. Info.* 2019, *16* (4), 2716–2725.

20. Gordon, W. J.; C. Catalini. Blockchain Technology for Healthcare: Facilitating the Transition to Patient-Driven Interoperability. *Comput. Struct. Biotechnol. J.* **2018,** *16*, 224–230.

21. Grover, J.; Garimella, R. M. Reliable and Fault-Tolerant IoT-Edge Architecture. In *IEEE Sensors*; New Delhi, India, 2018; pp 1–4.

22. Halonen, T.; Romero, J.; Melero, J., Eds. *GSM, GPRS and EDGE Performance: Evolution Towards 3G/UMTS*; John Wiley & Sons, 2004.

23. Jiang, W.;, Wang, Y.; Jiang, Y.; Chen, J.; Xu, X.; Tan, L. Research on Mobile Internet Mobile Agent System Dynamic Trust Model for Cloud Computing.

24. Kum, S. W.; J. Moon, T. Lim, and J. I. Park. A Novel Design of IoT Cloud Delegate Framework to Harmonize Cloud-Scale IoT Services. In *IEEE International Conference on Consumer Electronics (ICCE)*; Las Vegas, NV, USA, 2015; pp 247–248.

25. Lave, M.; Kleissl, J. Cloud Speed Impact on Solar Variability Scaling–Application to the Wavelet Variability Model. *Solar Energy* May **2013,** *91*, 11–21.

26. Lee, E. C.; Jung, H.; Kim, D. New Finger Biometric Method Using Near Infrared Imaging. *Sensors* 2011, *11* (3), 2319–2333.

27. Mehdi Khosrow-Pour, D. B. A. *Encyclopedia of Information Science and Technology*, 5th ed.; IGI Publications, USA, 2020.

28. Nagaraj, A. *Introduction to Sensors in IoT and Cloud Computing Applications*. Edited by Ambika; Bentham Science Publishers: Sharjah, 2021.

29. Nguyen, T. G.; T. V. Phan, B. T. Nguyen, C. So-In, Z. A. Baig, and S. Sanguanpong. Search: A Collaborative and Intelligent Nids Architecture for SDN-Based Cloud IoT Networks. *IEEE Access* **2019,** *7*, 107678–107694.

30. Peijina, H.; Tingb, J.; Yandongc, Z. Monitoring System of Soil Water Content Based on Zigbee Wireless Sensor Network. *Trans. Chinese Soc. Agric. Eng.* Jan 2011, *27* (4), 230–234.

31. Rasouli, M. R. An Architecture for IoT-Enabled Intelligent Process-Aware Cloud Production Platform: A Case Study in a Networked Cloud Clinical Laboratory. *Int. J. Prod. Res.* **2020,** *58* (12), 3765–3780.

32. S, A. I.; S, G. G. Cross Layer Approach For Detection and Prevention Of Sinkhole Attack Using A Mobile Agent. In *2nd International Conference on Communication and Electronics Systems (ICCES 2017)*, 2017; pp 359–365.

33. Sharma, P. K.; Chen, M. Y.; Park, J. H. A Software Defined Fog Node Based Distributed Blockchain Cloud Architecture for IoT. *IEEE Access* Sept 2017, *6,*, 115–124.

34. Shi, C.; Liu, J.; Liu, H.; Chen, Y. Smart User Authentication through Actuation of Daily Activities Leveraging WiFi-Enabled IoT. In *18th ACM International Symposium on Mobile Ad Hoc Networking and Computing*; ACM, 2017; p 5.

35. Singh, S. K.; Rathore, S.; Park, J. H. Block IoT Intelligence: A Blockchain-Enabled Intelligent IoT Architecture with Artificial Intelligence. *Future Gen. Comput. Syst.* 2020, *110*, 721–743.

36. Tao, F.; Y. Zuo, L. Da Xu, and L. Zhang. oT-Based Intelligent Perception and Access of Manufacturing Resource toward Cloud Manufacturing. *IEEE Trans. Indust. Info.* 2014, *10* (2), 1547–1557.

37. Verma, P.; S. K. Sood, and S. Kalra. Cloud-Centric IoT Based Student Healthcare Monitoring Framework. *J. Ambient Intell. Humanized Comput.* June **2018,** *9* (5), 1293–1309.

38. Wang, T.; Zhang, G.; Liu, A.; Bhuiyan, M. Z. A.; Jin, Q. A Secure IoT Service Architecture with an Efficient Balance Dynamics Based on Cloud and Edge Computing. *IEEE IoT J.* **2018,** *6* (3), 4831–4843.

39. Xiao, L. et al. Intelligent Architecture and Hybrid Model of Ground and Launch System for Advanced Launch Site. In *IEEE Aerospace Conference*; IEEE: Big Sky, MT, USA, 2019; pp 1–12.

40. Yang, Z.; Zhou, Q.; Lei, L.; Zheng, K.; Xiang, W. An IoT-Cloud Based Wearable ECG Monitoring System for Smart Healthcare. *J. Med. Syst.* Oct **2016,** *40* (12), 1–11.

CHAPTER 12

Systematic Literature Review of Search-Based Software Engineering Techniques for Code Modularization/ Remodularization

DIVYA SHARMA* and GANGA SHARMA

G. D. Goenka University, Gurugram, India

Corresponding author. E-mail divya.07sharma@gmail.com

ABSTRACT

Due to rigorous changes, the maintenance of software systems has become difficult and complex. As a result, these rigorous changes can degrade the quality of the software process. One of the extensively used techniques, software modularization can be used to address this issue that converts the software classes attained from software products. Although to enhance the software quality, this can be a challenging task that automatically remodularizes the process. There are many parameters involved in improving the quality of the software process by automatically remodularizing the software system. Some of them include semantic domain, upgrading the continuous changing uniformity, number of the software changes by using search-based methods. SBSE (Search-based software engineering) study aims to transfer software engineering issues from human-based search to machine-based search using a range of methods from metaheuristic search, evolutionary computing paradigms, and operations research. The

Computational Intelligence Applications for Software Engineering Problems. Parma Nand, PhD, Rakesh Nitin, PhD, Arun Prakash Agrawal, PhD & Vishal Jain, PhD (Eds.)

concept is to exploit humans' reliability and creativity, and machines instead of requiring people to conduct the more tedious, error-prone, and thus expensive elements of the engineering process. Researchers have suggested some methodologies to increase the accuracy of the machine learning classifier for the prediction of software defects by assembling some machine learning methods, using an algorithm to improve, select features, and optimize parameters in some classifiers.

12.1   INTRODUCTION

Search-based software engineering (SBSE) is a name provided to a workplace in which software engineering (SE) is applied to search-based optimization (SBO). The term "Search" is used in SBSE to refer to the use of metaheuristic SBO methods. SBSE is understood using operational research (OR) techniques and metaheuristic "search-based" techniques. SBSE represents a method for applying metaheuristic search methods like stimulated annealing (SA), genetic algorithm (GA), and taboo search to the problem of SE.[1,5]

A search strategy like random search, local search (e.g., tabu search, SA, and hill-climbing), evolutionary algorithms (e.g., evolutionary strategies, genetic programming (GP), and GA), particles warm optimization, and ant-colony-optimization can resolve a problem in the validation and verification domain or software testing. Other modern buzz words in software testing are near remodel-based testing, interaction testing, service-oriented architecture testing, real-time testing, prioritization of test cases,[32] and generation of entire data testing.

In this survey, we talk about the modularization and remodularization used in SBSE. Usually, most of the software system maintenance effort is dedicated to knowing the software system structure. This work is made easier by modularizing a structure that has less coupling and maximum cohesion, simplifying it, and analyzing the side effects of a change. Coupling is the extent of dependence between the modules and interdependence within a single module is interdependence. Low coupling is the indication of a well-designed software system.

The notion of software cohesion was described by the person who defined it as the degree to which a module's inner content is linked. In object-oriented software, cohesion is typically applied at the class level and may be expanded to the package level. It offers high reliability and maintenance when combined with high cohesion.[3,14,18,19]

We define an effective reorganization in the architecture of the system as remodularization with the main purpose to improve its internal consistency and therefore without adding fresh characteristics or bug fixation. In many cases, structural aspects, such as static dependencies between architectural entities, guide remodularization. For instance, a common recommendation is to optimize cohesion and to minimize coupling that can be followed manually, or with the assistance of semiautomatic remodularization instruments. Nevertheless, the reality is that there is no clear consensus that makes a healthy architecture. The subjective notion of architectural quality is difficult to measure with standard quality metrics. For example, recent work has begun to challenge the dogma for structural coupling or cohesion which states that "coupling and cohesion do not seem to be the dominant driving forces when it comes to modularization." Other research has shown that systemic cohesion metrics are normally divergent in the evaluation of the same refactoring actions. Semantic clustering is a strategy based on data collection and clustering methods to obtain sets of comparable classes according to their vocabulary in a scheme. Semantic clustering is a strategy based on data collection and clustering methods to obtain sets of comparable classes according to their vocabulary in a scheme.[4,14,26,33,49]

There are two categories of SE: first, Black Box optimization, a standard combinatorial optimization issue. Second, White Box issues in which source code activities need to be considered. SA, local search, GAs, and GP, Greedy Replant Algorithm, hill climbing (HC), linear programming (LP) methods, and integer linear programming are the most used optimization and search techniques.[1,7]

The first effort to optimize SE has been revealed by Miller and Spooner (1976). In 2001, Jones and Harman first used the SBSE term.

Two major ingredients for SBO use in SE problems were described by Jones and Harman: The fitness function the definition and choice of representation of a problem.[5,17] Software testing was introduced to SBSE, including automatic test case generation (test information), test case prioritization, and test case minimization. Certain attention has been paid to regression testing.

As software today is growing quickly in size and complexity, software reviews and testing are playing a major role in the software development process, particularly in software detection.

Compared to software testing and reviews, software defect prediction methods are much more cost-effective to identify software defects. We review search-based approaches used in the literature for the SEPM assignments covering the period between January 1992 and December 2015.

We pick and use the NSGA-II algorithm from the current heuristic search algorithms to discover a healthy modularization. We chose NSGA-II as amongst the most effective GAs as recommended in the literature and have been successfully used in a variety of software remodularization search-based solutions. We are introducing a novel strategy to identify the best sequence of refactoring activities that optimize the initial NSGA-II modularization.

The strategy utilizes measures of coupling (external dependencies package) and cohesion (inner dependencies package) to predict the performance enhancement.[43,47]

We suggest utilizing, to structural dependencies, a semantic metric to define conceptual dependencies between code components. We regard their factoring effort calculated with the measurement of achieved improvement (RRAI)[11,17] rate per refactoring as a minimizing goal.

We notice that this factor is considered by only a few existing methods and that the number of code modifications was used as an indication of a refactoring attempt in most of them.

- In the software field, a significant use of cluster analysis is for modularization of a software system by combining comparable or interrelated software entities to obtain maximum cohesion and minimum coupling.[13,19]
- Entities within a cluster have comparable features or characteristics and vary from entities in other clusters.
- As the software modifications and develops over the period, the undisciplined attitude to software maintenance will inevitably have an adverse impact on software quality.[12,18,32]
- The design of the scheme may eventually alter. Appropriate abstractions are required to comprehend and cluster the structure.[9,17,30,50]

- Architectural level opinions should be generated from the source code directly. In the source code, Bunch creates a graph of entities as well as relationships which are explained also in this document.
- The design extraction begins with the parsing of the source code for determining the elements and relationships in the source code. The parsed code is then evaluated to generate opinions of the software framework.

12.2   METHODS

Kitchenham[27,32,37] acknowledged three basic phases in the systematic review. These are as follows: planning, reporting, and conducting. The planning concerns the need for evaluation and the review protocol to be established. A review protocol in the planning phase describes explicitly the research issues that will be dealt with throughout the review.[12,18]

A review study aims to explore and analyze the literature to evaluate its existing state. In the mind of a researcher, the aims to direct the research approach and the research process should be evident in these questions. Research questions are based on the aims of the studies on a particular phenomenon or topic.[12,18,41]

Besides, a review protocol covers the search plan to extract appropriate evaluation studies, criteria for unequal selection of candidate studies, the qualitative analysis procedure for candidate studies, and the process of collecting as well as synthesized the information from taken primary studies to correctly address study issues.

A review phase includes executing the policies decided in the review protocol for collecting, extracting, monitoring, and synthesizing[7,10] the review studies.

Finally, the analysis results are reported in the reporting phase. The research questions must be developed with clear goals in mind to direct the research process and the research methodology.[10,18,42]

Research questions are focused on how we interpret a certain topic or phenomenon in studies. Also, a review protocol covers the search plan to extract appropriate evaluation studies, criteria for unequal selection of candidate studies, the qualitative analysis procedure for candidate studies, and the process of collecting as well as synthesized the information from taken primary studies to correctly address study issues.[7]

The goal of the review is to explore the use of various SBSE search methods in code modularization, as previously stated.

12.2.1 RESEARCH QUESTIONS

The work aims to research current methods in the SBSM field and to enhance the available methods to make the software refactoring process more efficient for a practical software development situation.[3,17,27,37]

An overall many-objective configuration is then utilized to incorporate the research in the tool to ensure that the software input can be preserved and enhanced in realistic terms across different properties, with minimal effort. The initial seven research questions referred to below are equal to the original research study.[11,32,48] The following research questions address these research objectives.

TABLE 12.1 Research Questions.

Research question no.	Research question
Q1.	What kind of methods improvement is prepared for software defect prediction?
Q2.	What are commonly used applications in SBSE (search-based software engineering techniques)?
Q3.	Which testing should be used for comparing the prediction capability of various search-based techniques for SBSE?

Source: Modified from Ref. [17]. With permission.

12.2.1.1 THE SEARCH PROCESS

Search processes are manual processes that are specific to conference proceedings and research articles since the year 2004.[12,17,18]

To design and develop a search strategy, we have developed and planned our search terms using the following steps to select appropriate studies in the literature.

- The research questions are fragmented into relevant logical units.
- Classify the search provisions from applicable abstracts, keywords, and paper titles.

- Establish the corresponding terms and synonyms for all the research studies.
- To accumulate and join synonyms and equivalent terms together, use Boolean OR.
- To accumulate and join main evident search terms, use Boolean AND.

Based on the above-mentioned steps, the following search string has been used to classify the set of relevant studies accessible in five major databases, namely: Information and Software Technology, IEEE Xplore, ACM Digital Library, Springer Link, Science Direct, Empirical SE Journal, IEEE Proceedings Software, ACM Computer Survey, and Wiley Online Library.

Keyword parameter, content, abstract as well as title are the basis of searches. The selected conferences, articles, and journals are referred to in Table 12.1.[13,16,32] These research studies were selected as they either incorporate the literature surveys or pragmatic studies.

These have been further used as sources for other SLRs[13,17] related to SBSE. We also looked manually for several well-known SBSE papers like Genetic and Evolutionary Computation Conference (GECCO) proceedings to make an exhaustive search.

Such manual and automatic attempts have led to a set of candidate research studies whose reference lists have been also scanned further to draw up more relevant studies. Whilst we used valuable insight from review studies[11,21,22] and removed them from the reference lists, these are not a part of the systemic review process since empirical results were not given.

12.2.1.2   CRITERIA OF EXCLUSION AND INCLUSION

The following topics included peer-reviewed articles and journals. These are as follows:

Meta-analysis: This included distinct individual and organizational research studies[7,17] to improve the overall effect which could further dissolve the abruptness.

All the ret articles, research papers, journals, and conference proceedings were included,[3,9,18,32] where literature review was the primary element.

The excluded research studies are as follows:

Informal literature reviews have no research questions, not properly outlined search process,[10,27] and no proper formal data extraction process.

Replica reports of the same research studies as distinct articles exist in different journals and conference proceedings[16] although the most relevant study was incorporated in this research review.

12.2.1.3 QUALITY ASSESSMENT

Each SLR[12,17,18,29] was evaluated and assessed. The criteria were based on these QA questions:

QA1: Are the exclusion and inclusion criteria reviews defined appropriately[12,15,36]?

QA2: Is this literature search covers all the appropriate facts and details[6,26]?

QA3: Are the basic details and research studies effectively described[11,26,28]?

QA4: Have the reviewers analyzed the quality parameters of the included research studies[2,17,50]?

The questions have been scored as follows:

QA1: Yes (Y), the inclusion criteria are explicitly called in the study, partly (P), the inclusion criteria are implicitly defined, No (N) the inclusion criteria[5,18] are not defined in the research study.

QA2:Y, researchers have searched for the relevant area of a topic in digital libraries including search strategies[3,6,8] and researchers have referenced all the articles, journals with the relevant area of interest; P, researcher have searched few digital libraries with no additional search schemes and they restricted themselves with a set of conference and journal proceedings; N, researchers[32,37] have searched no or minimal digital libraries with an extremely constrained set of conference proceedings.

QA3: Y, the data and knowledge presented is relevant to the research study[12]; P, past summary data of primary research has been provided[13]; N, researchers have not conducted any primary research on the relevant area.[4,32]

QA4: Y, the researchers have clearly defined all the quality parameters[12,17] and constraints from the primary research studies; P, there are quality and maintenance constraints[9,29,46] that are explicitly defined in the research questions; N, there has been no quality assessment conducted by the research body.

The quality extraction process was conducted based on the above quality assessment methodology. The scoring was $Y = 1$, $P = 0.5$, and $N = 0$ or undefined. Each research paper was analyzed and evaluated independently.

12.3 DATA COLLECTION

The data collected from each research study is as follows:

Sources and Process—The primary online sources used for information extraction are given below:

(http://dl.acm.org/) ACM digital library

(http://ieeexplore.org/)IEEE Xplore

(http://www.sciencedirect.com/) Science Direct

(http://www.springer.com/gp/) Springer LNCS

(http://scholar.google.co.in/) Google Scholar

The data collected was analyzed, checked, and extracted. Thus, ensuring quality assessment[6,26,33] of all the data collected from the primary research studies.

12.3.1 DATA ANALYSIS

The data collected was converted into a tabular form to conduct the following:

- The SLRs have referenced SBSE research papers along with the SLR guidelines[6,25,27,41] which addressesRQ1 and RQ2
- The number of literature reviews conducted and published each year[24,27,39,50] with its referenced sources. This also addresses RQ2 and RQ3.
- The research topics are covered as per the relevant area of interest by the SLRs[11,25,29,42,51] with its referenced scope. This addresses RQ2.
- The quality assessment of each SLR has been observed[12] and calculated which addresses RQ1.
- The research trends[15,27,38] and technical area covered addressing RQ3.

- The number of primary and secondary studies[1,18,37] were conducted in the research study referencing RQ5 and RQ6.
- The association of authors at institutional and organizational level addressing[21,24,27] RQ2.
- The results identifying the benefits and drawbacks of SBSE methods and modularization/remodularization focusing on RQ2 and RQ3.[18,30,39]

12.3.1.1   SLRS QUALITY EVALUATION

Quality assessment was done in Section 2.4 where quality was evaluated on various parameters and criteria by using different SLRs.

The quality scores[12,17,18] are given on various relevant parameters as described in Section 2.4 with Yes (Y), Partial (P), and No (N)[5,6,26] agreement from distinct researchers.

This enhances the quality of the research work and therefore various agreement and disagreed points can be collected and refined which should be helpful in future work.

12.4   RESEARCH QUESTION DESCRIPTION

12.4.1   DESCRIPTION OF QUESTION 1

What kind of methods improvement is prepared for software defect prediction?

Defect prediction is very important in software reliability and software quality. Compared to other software quality engineering studies, defect forecast is a novel area.[18,20] By covering essential predictors, the data type to be collected, and the role of the model of defect prediction[9,17] in software quality; it is possible to identify the interdependence between defects and predictors. This chapter provides with extensive ideas on the forecast of software defects and future study avenues.[38,48]

Preventive detection of software faults in software projects enables managers to take adequate decisions and planning restricted project resources in a more systematic and coordinated way.

In general, we should concentrate on different aspects of the problem.[8,18,25]

- Defect prevention
- Defect detection
- Defect analysis
- Defect correction

As defect prediction is a novel area, several prediction models suggested in this chapter will be discussed. Complexity and size metrics are used in the present prediction models to prevent any flaws that may happen during the project's operation or test stage. Reliability-based models utilize the operational profile of a system for forecasting failures that the project is going to face in another model of defect prediction (Fig. 12.1).

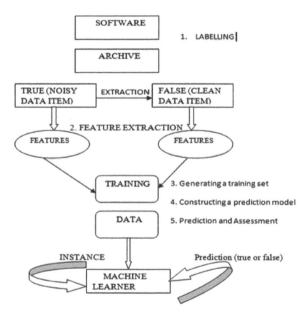

FIGURE 12.1 General defect prediction process.
Source: Modified from Ref. [18]. With permission.

Information gathered in testing and detection[12,18] of defects is also evaluated in most projects to assist in forecast defeats for comparable project kinds.

Since all defect prediction models have regions where they are brief, however, a search has been carried out for a model capable of predicting defects in a wide range of projects.[11,49] The multivariate model of defect prediction has been declared a model which may address this problem, but as of now no all-embracing model was discovered.[48]

With the significance of implementing the greatest quality rates in systems, improving methods for predicting defects has become essential so that at the early stage leading to the project, they can predict further defects.[17,31,40,49]

12.4.2 DESCRIPTION OF QUESTION 2

What are common applications used in SBSE?
SBO is only possible with two key ingredients for SE issues.[12,18] Choosing the problem representation. It lists the documents for each of them, drawing popular topics like search technique type used, the assessment nature, and the fitness definitions.

SBSE is a very appealing alternative because of its simplicity and willingness to use it. A software engineer is usually represented appropriately for their challenge as one cannot do more engineering without a way to reflect the challenge at hand. With these two components, SBO algorithms can be implemented.

These algorithms use distinct methods to find alternatives that are ideal or close to optimal.[17,41,45,50]

12.4.3 HILL CLIMBING

It begins with an original candidate solution randomly selected. The components of a set of "close neighbors" to the present solution are taken into consideration at each iteration. For every application to which HC is applied, what constitutes a "near neighbor" must be specified, because this is a particular issue. Usually, determining a near-neighborhood identification process is comparatively simple; other candidate solutions are near-neighbors, which are a "tiny mutation" away from the present solution. In every iteration of the key loop, the hill-climbing algorithm takes into consideration the set of near neighbors.[15] If the fitness of a neighbor is higher, a move to a fresh present alternative is made.

The move is to the first neighbor identified on the next climbing hill to have an improved fitness.[10,12]

The whole set of the neighborhood is examined in steep climbing hills to locate the neighbor that provides the biggest rise in fitness.[17,19]

12.4.3.1   SIMULATED ANNEALING

This may be considered as a hill-climbing variation that prevents the issue of local maxima by allowing moves to fewer fit people. The strategy provides more opportunities in the previous phases of search space exploration to consider fewer fit people.[3,15,18]

This opportunity is gradually decreased until the strategy in the very final phases of search space exploration becomes a traditional hill climb.

SA is a metallurgical annealing simulation in which an extremely heated metal can slowly decrease in temperature, thus improving its power. The atoms have less liberty of motion as the temperature reduces.

But in the previous (hotter) stages, the increased liberty allows the atoms to "explore" different energy states.[18,27,35,53]

12.4.3.2   GENETIC ALGORITHMS

This uses population and recombination ideas in GAs. GAs were the most used search technique in SBSE of all optimization algorithms, though this may have largely been for historical or sociological reasons,[10,15] rather than scientific or engineering reasons. That is, perhaps GAs, with their scientifically nostalgic rejection of Darwin, merely have a more natural attraction to scientists who, with a possibly confusing set of feasible SBO techniques, must choose one to start experiments on a fresh application area of SE.[7,18,26]

12.4.3.3   APPLICATIONS OF GENETIC ALGORITHMS IN SOFTWARE ENGINEERING

In several areas of software development, GAs was already used. In this chapter, we look at some reported results from the SE implementation of GAs.[19,25,32] Certainly, the list is not complete. It just acts as an indicator that individuals understand GAs' capability and start taking advantage of introducing them to software growth. In GA software metrics (design and coding)[16,30] are used as the numerical value associated with software development. Usually, metrics consist of the definition of the equation; however, this technique has been constrained by the fact that Vankudoth et al.[15,17] operate on a system software component choice, and all the interconnections of all the parameters are completely known. This is a crucial design phase choice and has a significant effect on the different quality characteristics of the system. To classify system software components based on architectural style, the software functionalities must be assigned among the software components.[14,18] The software functions must be assigned to the software components in selecting the architectural style. The author presents a GA-based technique that uses instances to define software components and their duties using the concept and design operation of the GA as methods (Fig. 12.2).

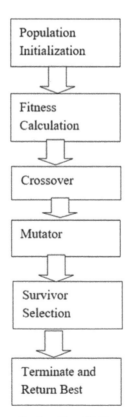

FIGURE 12.2   Flowchart of genetic algorithm (GA).

Modified from Ref. [23]. With Permissions.

The suggested technique is first to pick a suitable GA technique.[6,24,30] GAs also introduce strategies for optimization by simulating species evolution through natural choices.

Generally, the GA consists of two procedures. The first method is the selection of individuals for next-generation manufacturing and the second method is the manipulation of the chosen person by crossover and mutation methods to form the next generation.[31,35,41]

12.4.3.4   HOW GENETIC ALGORITHMS WILL WORK?

The GAs[5,16,25] begin with a big population in the beginning. Every person in that population constitutes a plausible alternative to the issue. The individuals are encrypted into a binary string known as a population chromosome.

Then the group of individuals starts competing to reproduce and develop the subsequent generation. Though, there is a function known as the fitness function that decides which competitors are entitled to reproduce.[18,25,37]

12.4.3.5   REPRODUCTION

This is done by swapping part of the code for two chromosomes to create new people. The crossover points (where the code bits are exchanged) are randomly chosen (for a simple algorithm variant). At this point, the chromosomes swap data to maintain the original data.[12,18,30]

12.4.3.6   MUTATION

This introduces variations in the next generation that prevent local minima from being reached. While crossover changes the genes between two chromosomes after a choosing crossover point randomly, the mutation picks one chromosome on the tree node and transfers the genetic content.[19,25,32,46]

12.4.3.7   ANT-COLONY-OPTIMIZATION-ALGORITHM

An ACOG varies from the algorithm provided by GP to improve performance. The primary loop runs for a user-specified iteration number.[13,17,27]
These are defined as the following:

- Initialization: set the system' initial parameters: states, variable, input, function, output, output trajectory, input trajectory.[12,18,52]
- Set value for initial pheromone trails.[18,32]
- Each ant should be placed individually with empty memory on the initial state.[12,16]

Due to the following reasons, the software does not meet the requirements while terminating conditions.

Construct ant solution: each ant creates a path by applying the transition function successively, depending on how attractive it can be to move from state to state and the trail level of movement.[12,18,41]

- Local Search is applied.[13,15]
- Best tour check: update, if there is an enhancement.[3,43]
- Update trails: on each route, evaporating a fixed share of the pheromone. Pheromone will be updated for each ant that is performing the "ant-cycle." Reinforcing the finest route with a set number of "elitist ants" performing the "ant-cycle."[11,14,39]
- Build a new population-based on pheromone paths by the following method. The operations shall be performed on the individual(s) with a probability of fitness chosen from the population[22,38]

12.5   DESCRIPTION OF QUESTION 3

What tests were used for comparing the predictive potential of various SBSE-based search techniques? The method used to measure the quality of the software is software testing.[13,16]

Quality is usually restricted to issues like completeness, accuracy, security, but it can also include nonfunctional specifications, including reliability, capability, performance, portability, maintenance, accessibility, and usability defined in the ISO Standard 91264.

The first domain to be discussed is software testing and the one that has undergone the most comprehensive studies in all fields of SE in which SBSE techniques[32,37] are used. The general concept behind these techniques to search-based test information generation is that search space is formed by the set of test instances (more commonly feasible program inputs) and the criterion of test capability is marked as a fitness function.[10,18]

For the search-based software testing sector, SBO techniques were first implemented.[23,27] The quality of the established software can be evaluated by software testing by implementing testing, the software bugs can be identified, although it does not provide free software for bugs.

Testing improves our trust in the reliability of software and helps to save time and cost of developing software.

Other testing areas in which SBSE is used include structural testing, regression testing, settings testing, integration testing, etc.[9,30,36]

To derive the optimum distributions of probability, the HC method has been introduced to statistical testing.[17,18]

SBST (Software-based self-testing): The approaches are efficient for periodic testing of deeply integrated processors in low-cost embedded systems for intermittent and permanent operative defects.[21,24,27] In comparison to hardware self-testing, SBST is nonintrusive because testing is done in standard circuit mode. Software-based testing allows the use of configurable random-pattern generation systems without the need for overhead testing[30,37] because of the programmability of software.

Also, software instructions can control test patterns through a complex processor, avoiding blocking test data because of nonfunctional control signals, such as for hardware-based BIST logic. The proposed technique benefits from deterministic structural testing by targeting the need for structural testing of less complicated parts. Because implementation of component test and reaction collection are conducted using directions rather than scan chains, no region or overhead output is required and at-speed the test application is conducted.[41,46]

More significantly, the Gigahertz processor can be tested at speed of low-speed testers by switching the function of internal testers from implementing testing to loading testing programs and unload answers.[22]

The SBST[23,27] offers low fault-detection latency for hardware as well as information-redundancy schemes but imposing considerable to massive hardware overhead.

It switches between performance overhead and defect detection latency, identifies both intermittent and permanent defects, and does not require immediate fault detection for low-cost no safety-critical embedded systems.[31,36]

Regression Testing—it is testing that a program has not regressed, that is, in the previous version the features that have worked still work in the current version. It is costly but an essential job to maintain a software system[14,16] or for software development.

Since software maintenance costs account for approximately 2/3 of the entire software, the regression testing must be taken more and more carefully by both researchers as well as project managers.

Optimization techniques—The SBO methods are utilized to address SE issues. These approaches are used during the entire SE model life cycle, for example, from requirement induction to maintenance of software. Several methods for search and optimization are available to assist solve SE issues. Many techniques of search and optimization help to solve the SE issues. These methods are divided into two categories: (1) metaheuristic search techniques, (2) classical technique.[18,26]

Classical techniques—LP is one of the approaches used to solve problems that clearly describe the single goal and the constraint set which could be represented as linear equations. It is mainly used for plant and resource allocation problems.[23,32]

Metaheuristic search techniques—The best solution is given by such methods, since they provide heuristic search region data, resulting in an almost ideal solution to the SE issues.

A brief description of the most used metaheuristic strategies—GA, SA, and HC.[25,27,29]

12.6   RESULT ANALYSIS

12.6.1   SEARCH RESULTS

Table 12.2 describes the search results of the search procedures during the research work conducted with the help of primary studies using different SLRs.

Table 12.2 elaborates the problem addressed with the relevant techniques, datasets, and parameters used. This has been analyzed and depicted author and year wise stating the algorithms. Meanwhile few research articles have been searched and defined by this search process.[35,37,54]

These research papers include articles, journals, conference proceedings from 2005[6,26,31] for SBSE for code modularization. A brief of each research proceeding has been described in the below mentioned tabular form.[39,42,45]

Table 12.2 elaborates RQ2 indicating the scope of the research studies that have been considered referring to the research trends and limitations for SBSE where code modularization results were compared with existing search-based techniques and tools.[15,18,21,36]

TABLE 12.2 Summary of Applications of SBSE Trends and Techniques (RQ2).

Authors	Years	Problems addressed	Techniques/ models applied	Datasets used	Comparison techniques
Simulated annealing (SA)					
Pai and Hong [4]	2006	Software reliability	SVM-SA model	Example from telemetry software system, AT&T Bell laboratories [5]	Sum of square errors, normalized root
Waeselynck et al. [6]	2006	Test generation, safety related property, validation, testing	SA landscape concept	Methods related to interfaces and API's	Diameter with true solution
Boukitif et al [7]	2006	Maximize OO model verification	SA & Bayesian Classifier	Quadric Algorithmic Complexity	Java sets
Pendharkar [9]	2010	Software error prediction complexity	NN exhaustive algorithm	Classes from Promise [6] functions	Recall
Genetic algorithm (GA)					
Millar et al [10]	2006	Automatic the test scenarios in java classes	Data coverage, GA	Java API framework	Datasets and fitness function
Srivastava [12]	2009	Verifying software quality	GA	Fitness criteria	Minimum and maximum algorithms
Babamir et al [14]	2010	Construction of graphical format for increasing quality of software	Data set generation for test	Code generated numbers	Paths related to targets and fitness function

TABLE 12.2   *(Continued)*

| Rao et al. [19] | 2013 | Validating software testing | GA | | Missionary technique | Branch and code coverage |
| Puri et al. [20] | 2014 | Data set optimization | Control graphs | Java framework | | Test data generator |

Source: Modified from Ref. [29]. With permission.

Few research studies have been conducted under peer review research methodology.[12,16,18,33,36]

Table 12.2 describes the search results of the search procedures during the research work conducted with the help of primary studies using different SLRs.

Paper Published. Figure 12.3 demonstrates the number of articles that have been released each year since 1999. The biggest number of articles released in a year is four from 2004 to 2008. There are no publications released in 2009 and 2010, but there has been an increase in search-based refactoring studies since 2011.

There are at least four papers released a year between 2011 and 2018. The year 2012 was one of the productive years for search-based refactoring studies with eight publications released this year.

Years 2004, 2005, 2009, and 2010 are remarkable since no search-based refactoring papers have been issued. Overall, an average of three articles is released annually from 2004 to 2016.

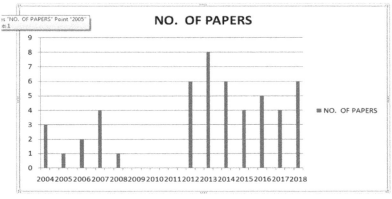

FIGURE 12.3   Each year number of papers published.

Paper Types Figure 12.4 shows the various paper kinds analyzed in the literature. Not all but one of the articles was released in newspapers or included in meetings. The other was included as a book chapter (Koc et al., 2012). Most of the articles in the journal and conference are from meetings. Table 12.2 lists meetings from which at least two of the articles analyzed originate[29] (Pizzi and Vivanco, 2004; Seng, 2006; Cinnéide and O'Keeffe, 2007a; Ouni et al., 2013; Mkaouer, 2014). This conference, which mainly concerns EAs, includes more articles as compared to maintenance conferences (ICSM, SANER, and CSMR) as well as SBSE also (SSBSE).[30,34,36]

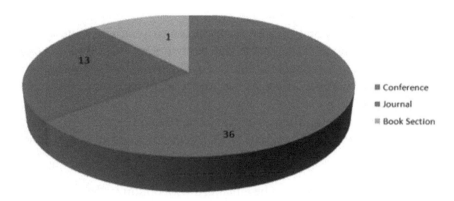

FIGURE 12.4   Types of paper analyzed.

12.6.2   LIMITATIONS

The techniques and processes mentioned in the present analysis have somewhat deviated from Kitchenham's guidelines[6,24,27,32]
    These are as follows:

- A manual search procedure was used for a set of articles, conference proceedings, as well as journals whereas an automatic search technique was not used.
- Data collection and data extraction should have been an automated process.

- Few SLRs analyzed and evaluated the quality of the primary research studies.
- Few SLRs include practitioner guidelines.

There can be data aggregation as well as data extraction issues when it comes to a detailed view of the systematic literature review. Meanwhile, the quality assessment criteria were the most complex criteria[12,15,25,33,54] for data assessment as it is somewhere subjective. However, quality criteria were calculated independently without any interference which should reduce the errors in the software program.[48,55]

12.7 CONCLUSION

This article offers an analysis of the various forms of search methods are used in SE techniques. The article demonstrates an overview of GAs, simulated annealing, modularizations, remodularization techniques, and algorithms used. We take different years and types of papers for techniques and algorithms.

After reviewing the discussions on each research issue, we suggest future work for investigators who wish to employ code modularization search methods:

- Speed and code size can be enhanced with optimization techniques. It must be decided when optimization is to be used since premature optimization is the core cause.
- Optimization must be implemented under reasonable conditions otherwise the readability of the code does not impede efficiency.
- Open problems have been analyzed such as package structure improvements, code changes, semantic domain, and consistency issues.
- A variety of other validation strategies, including cross-project validation, inter-release validation, or temporary validation, should be implemented in future work, reducing the validation bias and testing the models created.
- In addition to predictive precision, more work on assessment dimensions of search-based techniques should be focused on. This

may involve the cost efficiency, comprehensibility, and generalization capacity of the established models.

KEYWORDS

- **SBSE (search-based software engineering)**
- **modularization**
- **remodularization**
- **optimization**
- **metaheuristic algorithms**

REFERENCES

1. International Journal of Computer Applications (0975–8887) Innovations in Computing and Information Technology (Cognition 2015) Search Based Software Engineering Techniques. Arushi Jain M. Tech Student ITM University, Gurgaon, Aman Jatain Assistant Professor Amity University, Gurgaon.
2. Bradbury, J. S.; Kelk, D.; Green, M. Effectively Using Search-Based Software Engineering Techniques within Model Checking and Its Applications. In *Proceedings of the 1st International Workshop on Combining Modelling and Search-Based Software Engineering*; IEEE Press, May 2013; pp 67–70.
3. Räihä, O. A Survey on Search-Based Software Design. *Comput. Sci. Rev.* **2010**, *4* (4), 203–249.
4. Santos, G.; Valente, M. T.; Anquetil, N. Remodularization Analysis Using Semantic Clustering. In *2014 Software Evolution Week-IEEE Conference on Software Maintenance, Reengineering, and Reverse Engineering (CSMR-WCRE)*; IEEE, February 2014; pp 224–233.
5. Harman, M.; Jones, B. F. Search-Based Software Engineering. *Inf. Softw. Technol.* December **2001**, *43* (14), 833–839.
6. Mahouachi, R. Search-Based Cost-Effective Software Remodularization. *J. Comput. Sci. Technol.* **2018**, *33* (6), 1320–1336.
7. Kitchenham, B. A. Guidelines for Performing Systematic Literature Review in Software Engineering, Technical report EBSE-2007-001, UK, 2007.
8. Rawat, M. S.; Dubey, S. K. Software Defect Prediction Models for Quality Improvement: A Literature Study. *Int. J. Comput. Sci. Issues* **2012**, *9* (5), 288.
9. Ghannem, A.; Kessentini, M.; Hamdi, M. S.; El Boussaidi, G. Model Refactoring by Example: A Multi-Objective Search Based Software Engineering Approach. *J. Softw.: Evol. Process* **2018**, *30* (4), e1916.
10. Dumais, S. T. Latent Semantic Analysis. *Ann. Rev. Inf. Sci. Technol.* **2004**, *38* (1), 188–223.

11. Blei, D. M.; Ng, A. Y.; Jordan, M. I. Latent Dirichlet Allocation. *J. Mach. Learn. Res.* **2003,** *3,* 993–1022.

12. Zhou, J.; Zhang, H.; Lo, D. Where Should the Bugs Be Fixed? More Accurate Information Retrieval-based Bug Localization Based on Bug Reports. In *IEEE,* 2012; pp 14–22.

13. Artzi, S.; Dolby, J.; Tip, F.; Pistoia, M. Fault Localization for Dynamic Web Applications. *IEEE Trans. Softw. Eng.* **2011,** *38* (2), 314–335.

14. Jones, J. A.; Harrold, M. J. Empirical Evaluation of the Tarantula Automatic Fault-Localization Technique. In *Proceedings of the IEEE/ACM International Conference on Automated Software Engineering,* 2005; pp 273–282.

15. Jones, J. A.; Harrold, M. J.; Stasko, J. Visualization of Test Information to Assist Fault Localization. In *Proceedings of the International Conference on Software Engineering,* 2002; pp 467–477.

16. Abreu, R.; Zoeteweij, P.; van Gemund, A. J. C. An Evaluation of Similarity Coefficients for Software Fault Localization. In *Proceedings of the 12th Pacific RIM International Symposium* Dependable Computing, 2006; pp 39–46.

17. Harman, M.; Mansouri, S. A.; Zhang, Y. Search based Software Engineering: A Comprehensive Analysis and Review of Trends Techniques and Applications. Department of Computer Science, King's College London, Tech. Rep. TR-09-03, 2009; p 23.

18. Harman, M. Search Based Software Engineering for Program Comprehension. In *15th IEEE International Conference on Program Comprehension (ICPC'07)*; IEEE, June 2007; pp 3–13.

19. Rawat, M. S.; Dubey, S. K. Software Defect Prediction Models for Quality Improvement: A Literature Study. *Int. J. Comput. Sci. Issues* **2012,** *9* (5), 288.

20. Ducasse, S.; Pollet, D.; Suen, M.; Abdeen, H.; Alloui, I. Ackage Surface Blueprints: Visually Supporting the Understanding of Package Relationships. In *Proceedings of International Conference on Software Maintenance*; Paris, France, 2007.

21. Dorigo, M.; Stützle, T. The Ant Colony Optimization Metaheuristic: Algorithms, Applications, and Advances. In *Handbook of Metaheuristics*; Springer: Boston, MA, 2003; pp 250–285.

22. Chen, L.; Dey, S. Software-based Self-testing Methodology for Processor Cores. *IEEE Trans. Comput. Aided Des. Integ. Circ. Syst.* **2001,** *20* (3), 369–380.

23. Harman, M. The Current State and Future of Search Based Software Engineering.

24. Arcuri, A. Automatic Software Generation and Improvement Through Search Based Techniques (Doctoral dissertation, University of Birmingham), 2009.

25. Johnson, D. S.; Papadimtriou, C. H.; Yannakakis, M. How Easy is Local Search? *J. Comput. Syst. Sci.* **1988,** *37* (1), 79–100.

26. Korel, B. Automated Software Test Data Generation. *IEEE Trans. Softw. Eng.* **1990,** 870–879.

27. Moscato, P. On Evolution, Search, Optimization, Genetic Algorithms and Martial Arts: Towards Memetic Algorithms. Caltech Concurrent Computation Program, C3P Report 826, 1989.

28. Poli, R.; Langdon, W. B.; McPhee, N. F. A Field Guide to Genetic Programming. Published via http://lulu.com and freely available at http://www.gp-field-guide.org. uk, 2008.

29. Mohan, M.; Greer, D. A Survey of Search-Based Refactoring for Software Maintenance. *J. Softw. Eng. Res. Dev.* **2018,** *6* (1), 3.

30. Ahuja, S. P. A Genetic Algorithm Perspective to Distributed Systems Design. In *Proceedings of the Southeastcon 2000,* 2000; pp 83–90.

31. Amoui, M.; Mirarab, S.; Ansari, S.; Lucas, C. A Genetic Algorithm Approach to Design Evolution Using Design Pattern Transformation. *Int. J. Inf. Technol. Intell. Comput.*), June/August **2006,** *1* (1, 2), 235–245.

32. Antoniol, G.; Di Penta, M.; Neteler, M. Moving to Smaller Libraries via Clustering and Genetic Algorithms. In *Proceedings of the Seventh European Conference on Software Maintenance and Reengineering (CSMR'03),* 2003; pp 307–316.

33. Bass, L.; Clements, P.; Kazman, R. *Software Architecture in Practice*; Addison-Wesley, 1998.

34. Bodhuin, T.; Di Penta, M.; Troiano, L. A Search-Based Approach for Dynamically Re-Packaging Downloadable Applications. In *Proceedings of the Conference of the Center for Advanced Studies on Collaborative Research (CASCON'07),* 2007; pp 27–41.

35. Bouktif, S.; Kégl, B.; Sahraoui, H. Combining Software Quality Predictive Models: An Evolutionary Approach. In *Proceedings of the International Conference on Software Maintenance (ICSM'02),* 2002.

36. Bouktif, S.; Azar, D.; Sahraoui, H.; Kégl, B.; Precup, D. Improving Rule Set Based Software Quality Prediction: A Genetic Algorithm-Based Approach. *J. Object Technol.* April **2004,** *3* (4), 227–241.

37. Bouktif, S.; Sahraoui, H.; Antoniol, G. Simulated Annealing for Improving Software Quality Prediction. In *Proceedings of the Genetic and Evolutionary Computation Conference (GECCO 2006),* 2006; pp 1893–1900.

38. Bowman, M.; Briand, L. C.; Labiche, Y. Solving the Class Responsibility Assignment Problem in Object-Oriented Analysis with Multi-Objective Genetic Algorithms, Technical report SCE-07-02, Carleton University, 2008.

39. Briand, L.; Wüst, J.; Daly, J.; Porter, V. Exploring the Relationships between Design Measures and Software Quality in Object Oriented Systems. *J. Syst. Softw.* **2000,** *51,* 245–273.

40. Bui, T. N.; Andmoon, B. R. Genetic Algorithm and Graph Partitioning. *IEEE Trans. Comput.* July **1996,** *45* (7), 841–855.

41. Canfora, G.; Di Penta, M.; Esposito, R.; Villani, M. L. An Approach for qoS-aware Service Composition Based on Genetic Algorithms. In *Proceedings of the Genetic and Evolutionary Computation Conference (GECCO) 2005,* June 2005; pp 1069–1075.

42. Canfora, G.; Di Penta, M.; Esposito, R.; Villani, M. L. QoS-aware Replanning of Composite Web Services. In *Proceedings of IEEE International Conference on Web Services (ICWS'05) 2005,* 2005; pp 121–129.

43. Canfora, G.; Di Penta, M.; Esposito, R.; Villani, M. L. A Lightweight Approach for QoS-aware Service Composition. In *Proceedings of the ICSOC 2004–short Papers*; IBM Technical Report, New York, USA, 2004.

44. Cao, L.; Li, M.; Cao, J. Cost-driven Web Service Selection Using Genetic Algorithm. In *LNCS 3828,* 2005; pp 906–915.

45. Cao, L.; Cao, J.; Li, M. Genetic Algorithm Utilized in Cost-Reduction Driven Web Service Selection. In *LNCS 3802,* 2005; pp 679–686.

46. Che, Y.; Wang, Z.; Li, X. Optimization Parameter Selection by Means of Limited Execution and Genetic Algorithms. In *APPT 2003, LNCS 2834*, 2003; pp 226–235.

47. Chidamber, S. R.; Kemerer, C. F. A Metrics Suite for Object Oriented Design. *IEEE Trans. Softw. Eng.* **1994,** *20* (6), 476–492.

48. Clarke, J.; Dolado, J. J.; Harman, M.; Hierons, R. M.; Jones, B.; Lumkin, M.; Mitchell, B.; Mancoridis, S.; Rees, K.; Roper, M.; Shepperd, M. Reformulating Software Engineering as a Search Problem. *IEE Proc. Softw.* **2003,** *150* (3), 161–175.

49. Deb, K. Evolutionary Algorithms for Multicriterion Optimization in Engineering Design. In *Proc. Evolutionary Algorithms in Engineering and Computer Science (EUROGEN'99)*, 1999; pp 135–161.

50. Dipenta, M.; Neteler, M.; Antoniol, G.; Merlo, E. A Language-Independent Software Renovation Framework. *J. Syst. Softw.* **2005,** *77*, 225–240.

51. Doval, D.; Mancoridis, S.; Mitchell, B. S. Automatic Clustering of Software Systems Using a Genetic Algorithm. In *Proceedings of the Software Technology and Engineering Practice*, 1999; pp 73–82; Falkenaur, E. *Genetic Algorithms and Grouping Problems*; Wiley, 1998.

52. Fatiregun, D.; Harman, M.; Hierons, R. Evolving Transformation Sequences Using Genetic Algorithms. In *Proceedings of the 4th International Workshop on Source Code Analysis and Manipulation (SCAM 04)*, September 2004, IEEE Computer Society Press; pp 65–74.

53. Gold, N.; Harman, M.; Li, Z.; Mahdavi, K. A Search Based Approach to Overlapping Concept Boundaries. In *Proceedings of the 22nd International Conference on Software Maintenance (ICSM 06)*, USA, September 2006; pp 310–319.

54. Goldsby, H.; Chang, B. H. C. Avida-mde: A Digital Evolution Approach to Generating Models of Adaptive Software Behavior. In *Proceedings of the Genetic Evolutionary Computation Conference (GECCO 2008)*, 2008; pp 1751–1758.

55. Goldsby, H.; Chang, B. H. C.; Mckinley, P. K.; Knoester, D.; Ofria, C. A. Digital Evolution of Behavioral Models for Autonomic Systems. In *Proceedings of 2008 International Conference on Autonomic Computing*, 2008; pp 87–96.

CHAPTER 13

Automation of Framework Using DevOps Model to Deliver DDE Software

ISHWARAPPA KALBANDI[1*] and MOHANA[2]

[1]*Computer Engineering, Dr. D.Y. Patil Institute of Engineering, Management & Research Akurdi, Pune, India*

[2]*Electronics and Telecommunication Engineering, RV College of Engineering, Bangalore 560059, India*

Corresponding author. E-mail: ishwar.kalbandi@gmail.com

ABSTRACT

The increasing need for faster delivery of software product and need to innovate in delivery basis, Development and Operations (DevOps) is formulated to overcome shortcomings of agile system of software delivery model. So to overcome from these problems, the strategy of software development needs to be done. The system to be developed with proper planning in such a way that it should take less time for code deployment by software developer and sprints of the development should be short. DevOps model proposes to break the silos between development and operations. Strategy for Dynamic Diameter Engine (DDE) software positions networks to handle and adopt networks to DevOps-model-based software delivery. Automation of test execution cases and development of Continuous Integration Framework, this strategy helps in more interaction between development and quality in enhancing quality of software produced. Currently, an automation framework Rammbock is being setup

Computational Intelligence Applications for Software Engineering Problems. Parma Nand, PhD, Rakesh Nitin, PhD, Arun Prakash Agrawal, PhD & Vishal Jain, PhD (Eds.)

to run DDE functional test cases. Automation of functional test cases has started and target for 100% automation. Automation framework is added with tools that help in validation and debugging the test cases which are triggered automatically. Counters tool developed using ROBOT Integrated Development Environment validates counter values obtained in DDE during execution of test cases. Tracing tool sets trace level automatically during execution of test cases so that log files can contain enough information to debug in event of test case failures. Development of automation framework for counters and tracing has resulted in an optimized testing. Counters tool is developed using Robot Framework, which is operating system independent, making it more flexible to adopt. Counter tool has made the task easier by providing instant output rather than looking graph/XML file manually. Tracing tool makes user to enable tracing quickly, the debugging issue can be solved just by taking inputs in short time. These tools help in a complete automation of test cases, thereby avoiding manual testing, which is prone to errors and time consuming.

13.1   INTRODUCTION

Communication between core elements of 3G or 4G network is essential for authenticating, accounting, and authorizing subscribers to wireless services of voice and data. Diameter is the protocol predominantly used in Long Term Evolution (LTE) and 3G networks for signaling, authorization, authentication, and accounting information. Following are major nodes of LTE network.[17]

- Mobility Management Entity (MME)
- Home Subscriber Server (HSS)
- Serving Gateway (SGW)
- Packet Data Network-Gateway (PDN-GW)
- Policy and Charging Control Entity/Function (PCRF)

Diameter enables communication between these nodes to handle various events including attach, detach and tracking area updates of subscribers that may or may not involve in change of MME/SGW, policy modifications, and charging updates. Diameter is an Authentication Authorization and Accounting (AAA) protocol, it is an extensible control plane message exchange protocol designed as a successor protocol to Remote Authentication Dial In User Service (RADIUS). With reference to Open Systems Interconnection (OSI) layered model it comes in application layer. A protocol that is message based is called as diameter in which diameter signaling nodes exchange messages and receive acknowledgments between nodes for every message. At transport layer it uses transmission control protocol TCP/IP or Stream Control Transmission Protocol (SCTP), secured as underlying transport protocol with Transport Layer Security (TLS) or Datagram Transport Layer Security (DTLS) as secure variants which makes diameter reliable. Diameter is the successor of RADIUS, RADIUS is a AAA protocol based on User Datagram Protocol (UDP). TCP uses hand-shaking mechanism (implicit) for providing reliability, ordering, and data integrity, unlike UDP which does not have these features. Unreliability and small message size are disadvantages of Radius. Some elements in network (3G, 4G) are only reachable via intermediate elements. Such intermediate elements are called diameter agents or diameter routing agents or diameter signaling controllers. Routes are used to determine how to get from a client to an indirectly reachable server. A diameter agent which handles relay, proxy, redirect and interworking, in native or virtual deployments, to reduce the network diameter routing many identical features has been used. With increasing traffic in LTE network due to more and more subscriber devices being connected to Internet, a reliable protocol with seamless routing capabilities is required. Diameter protocol which has evolved from RADIUS is promising in all IP networks of LTE. The chapter organization is as follows. Section 13.2 describes review of literature. Section 13.3 elaborates background or preamble of proposed work. Section 13.4 presents design and specifications of automation tool. Section 13.5 describes the implementation of the proposed work. In Section 13.6 detailed results are presented and Section 13.7 presents concluding remarks and future direction.

13.2 LITERATURE REVIEW

The diameter protocol is introduced because to enable more secure transaction, they use authentication and authorization process for security and access control are difficult than traditional. Because of these days a wide variety of recent technologies and various applications like wireless networks, mobile IPs, were used. There are existing protocols like RADIUS which fails to adjust with these kinds of requirements so need to design a protocol in such a way that it should satisfy control features of access keeping in mind that it can be extended further with flexibility.

Automation framework is enhanced using DevOps delivery model.[2,3] DevOps model is chosen because traditional waterfall model has the following challenges:

- This model takes high time for deployment and development. So work pressure will be increased and takes huge time for code deployment.
- Moreover, this model does not satisfy all operations completely. Due to these four challenges, that is, the production environment uptime 100% to be maintained is difficult. Usage of in affective tools infrastructure, the time complexity is increased due to monitoring number of severs, the product issue diagnosis and providing feedback were tedious task.[4,5]

Hence a software development strategy is required to overcome these problems, from developer's point of view a system which enables code deployment without any delay, a system where work happens on current code itself, that is, development sprints are short and well planned. The system should be developed in such a way that it should have minimum 99% uptime in view of operations like administrations activity should be easy and monitoring the activity very effectively and providing the system feedback. In addition, better collaboration between development and operations is most common requirement of operation team as well as developers. So in order to improve the productivity and collaboration work operation team and developers need to be combined in DevOps.[7,15]

In DevOps, single integrated group has the end-to-end responsibility of application (software), Integrated group consists of system admin, QA's, Testers, Developers, etc., so these people gather the requirements from the user to develop an application and testing, deployment and infrastructure development and finally monitoring and taking feedback from the end user and again implementing suggested changes.

Testing time can be reduced from days to hours in automated software testing, while it reduces repetitive tests. A time saving directly affects on cost savings, while less time taken applications cost will be very less.

To improve quality of the software, the testing scope and depth should be increased and this can be achieved by automated software testing by executing complex test cases during every test run. DevOps in software development provides quicker mitigation of software defects, better resource management, and stable operating environment.

13.3 PREAMBLE

The evolved core system of packet is LTE which is an access part of it. It was introduced in 3GPP R8. LTE network is proposed to meet requirements of flexibility in bandwidth and frequency, high spectral efficiency, short round trip times, and high peak data rates. Need for LTE

- To ensure end-to-end network quality of service.
- To develop a nonhierarchical system.
- To a complete packet switched network.
- To increased base of user and achieve higher data rate demands.
- To reduce transit time for user packets.

Mobile telecommunication world evolved from (1G) analog networks to (2G) Global System for Mobile Communication (GSM) network developed by European Telecommunication Standard Institute (ETSI).[11] 3G Universal Mobile Telecommunications System (UMTS) standard is third-generation cellular network evolved and LTE evolved over UMTS for all IP networks. LTE as a wireless network has improved spectral efficiency by implementing OFDMA and in combination with features;

- High data rates up to 75 Mbps (uplink) and 300 Mbps (downlink).
- Higher order modulation (32, 64 QAM)
- Large bandwidths (up to 20 MHz).
- Spatial multiplexing in downlink.

Structure of an LTE network can be broken down into multiple elements that include many nodes and interfaces. This starts at with User Equipment (UE), this process explains how UE receives an Internet connection from Packet Data Network (PDN) by utilizing LTE network. Concept of "bearers" that "route IP traffic from a gateway in PDN to the UE". Bearers contain information that defines Quality of Service (QoS) that user will receive from network. This is determined much later in core or applications section.

LTE basic nodes and interfaces

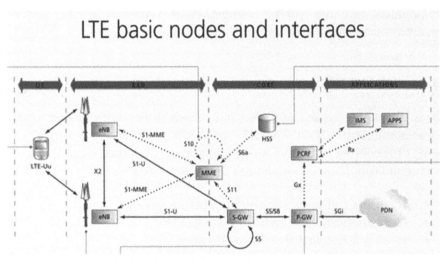

FIGURE 13.1   LTE basic nodes and interfaces.[19]

Structure of an LTE network is broken down into its nodes and interfaces as shown in Figure 13.1. These are mainly fragmented into four sections: UE, Radio Access Network (RAN), Core, and Applications. Firstly, UE is device that will connect to network. The UE will contain a Universal Subscriber Identity Module holds authentication information for device such as phone number, home network identity, etc. This is the information of phone's SIM card; UE monitors radio and conveys performance to evolved node B (eNodeB) channel quality indicator (CQI). To support the downlink and uplink of air interface UE is used. UE connects to network via physical nodes, eNodeBs utilizing AS protocols. The eNodeBs, in RAN section, are physical connection from UE into network. Network on eNodeBs are connected using X2 interface, this allows eNodeBs to be geologically spread out to allow UEs to connect in different locations. eNodeBs functions consist of mainly radio functions, which includes: To provide allocation of resource in UE radio admission control, mobility connection control, radio bearer control, and scheduling downlink or uplink, these functions are done by radio resource management. They also control IP header compression and ciphering of user data stream; all data sent over nodes must be encrypted. Since it is a connection between UE and MME, it should have ability to handle along with MME selection. eNodeBs also forward uplink data to Serving Gateway (SGW) from UE. Then there is the core, which consists of overall control of entire network. This consists of five main logical nodes: PDN Gateway (PGW), SGW, MME, PCRF, and HSS. The MME has many core functionalities needed for network. MME controls a lot of bearer functionality such as Non-Access Stratum (NAS) signaling for attachment and bearer setup/deletion. It is also in charge of bearer management functions including dedicated bearer establishment from UE to network. NAS signaling security is also handled here. MME also handles selection for handoffs, roaming, authentication, and a disaster warning message transmission system. HSS is utilized by MME to retrieve subscriber data since HSS is just a storage of subscriber data. This subscriber data includes Enhanced Presence Service (EPS) QoS subscriber profile and roaming restrictions list. There is also a list of accessible Access Point Names (APNs) which are access points to PDNs for UE to connect. HSS is also responsible for holding address of the current serving MME. Authentication vectors and security keys for the UE are stored here. The SGW is essential in handover of IP Packets to eNodeB. It is local mobility anchor that holds data bearers while MME determines which eNodeB the UE is connected to. This gateway also collects some information for charging to pass onto the PCRF later. It collects information about uplink and downlink charging per UE, PDN, and QoS class identifier. The PDN Gateway is connection to the Internet (PDN). P-GW enforces the downlink rate based on the QoS according to the rules from the PCRF. As well, as enforces uplink/downlink service level charging, gating and rate. Also, filtering packets are based on the QoS. The PCRF is where policy and charging functionality happens. It receives the UE information as well as the usages and, based on the rules' functions, it conveys the policy decisions to the PDN. Policy determines what QoS the user will get. Also based on the users' subscriptions, charging for the usages will be determined. Applications section is versatile. The main application possibility is IP Multimedia Subsystem (IMS). This introduces IP to allow packet switching instead of circuit switching, which voice calls used. This allows voice, text, and multimedia services to be connected over all networks since they can now all use IP.

13.3.1   CHARGING POLICY

The charging system is mainly classified into two categories:

- On-line Charging (OCS) and
- Off-line Charging System (OfCS).

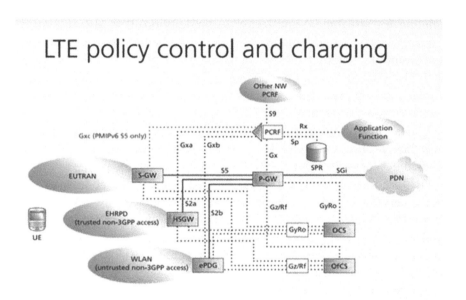

FIGURE 13.2   LTE policy control and charging.[20]

Figure 13.2 shows the interfaces of GY/Ro which is utilized in OCS. It only focus on Session Charging with Unit Reservation (SCUR) to explain how charging works because it splits up charging into smaller parts. P-GW sends the OCS a Credit Control Request (CCR) Diameter Message and will receive a Credit Control Answer (CCA) back with granted units, services, etc. OCS will also create a session on the CCR Initial (CCR-I) request to allow for CCR Update (CCR-U) requests that update used units and requests more. This session will continue until CCR Termination (CCR-T) request is received to terminate session. The OCS can be split up into a few different components:

- *Call Control:* interfaces between P-GW and other parts of system's functionality. This can handle multiple different protocols but this example will focus on Diameter.
- *Subscriber Model:* Device/User related functionality such as lookup or update.
- *Service Model:* Core charging related entities (Bundle, Tariff, Bucket, etc.) and associated functionality.
- *Session Manager:* Session operations (create, update, delete, etc.) and associated functionality.
- *Charging Engine:* Core charging functionality.

It starts with the PGW sending OCS a CCR-I attachment. Call control receives that request and starts to handle it. It then gets the device info from subscriber model. Call control builds a session object and sends those two things along with a diameter message to charging engine. Charging engine gets subscription details from service model for the device. It then creates a reservation of units and builds a response object with granted units and a session object created by session manager. Session objects are used inside charging engine but session gets created or updated during call control functionality. Response object is handled by call control and a session is created. Finally, it returns a CCA with granted units to P-GW. P-GW can update session with CCR-U requests. If user is using the device, session will continue to be updated until session is terminated. CCR-U is sent to call control, device and sessions are retrieved from subscriber model and session manager respectively. A session object is built again. A request object along with device, session object, and diameter message are sent to charging engine. Charging engine has another step this time though it gets subscription details from service model and applies conditions and executes rates to determine tariffs according to subscription's bundles and units used by device. Bundles contain information about rates at which device will be charged for units it uses. More units are then reserved. Response object is created with new granted units and an updated session object. Response object is handled by call control and session is updated. Again, a CCA is returned to PGW with a success result code. CCR-T will not grant or reserve new units as like in CCR-U. Session is terminated here which stops any new updates. To start a new session, a CCR-I will have to be sent, which will restart this whole process.

Devices have subscriptions to bundles that contain a few bits of information. Firstly, a numeric priority value which is used to determine which bundle applies when there are multiple bundles for device's subscription. There is an applicability condition, which determines which bucket is used depending on criteria, for example, time of day (a bundle might have discounted evening rates so user is not charged as much for evening usage). The bucket will have information about the quota and quota type, for example, a bundle might allow for 300 MB of data for free then, after that is used, it will be charged at $0.10 per 1 MB. Same device can be charged at multiple different rates for different quota types, different dates, or at different levels of usages.

There are many different attributes that are looked at when determining the rate; this is difficulty of charging. All these must be considered every time charging happens and it must be done incredibly quickly because there are hundreds of thousands of devices that are using network whose usage must be tracked and charged according to their subscription. Buckets keep track of usage and rate will depend on bucket. Rate is how much the device is charged for usage. Rate depends on all bundle information. Based on the usage, the charges will be deducted by the device. Along with SCUR, there is also Immediate Event Charging (IEC) and Event Charging with Unit Reservation (ECUR). SCUR utilizes initial, update, and termination request types. IEC only uses one event request that causes immediate charging based on that event request. ECUR is in between with an initial and termination request. ECUR and SCUR reserve units before the service is delivered whereas IEC just sends one request without need for unit reservation.

Performance is incredibly important within charging section. There can be hundreds of thousands of active sessions at any time. Each of these sessions constantly receives CCRs which need to be handled quickly and correctly. Mistakes can be very costly to network owner with outages or any slowness upsetting the customer. Every aspect of charging must be done quickly. Call control must collect, handle, and pass on information to charging engine efficiently. Charging engine must determine rates and tariffs as fast as CCR-Us are received. This also means that any database query must be fast so database choice is a key step in creating a good charging service. Ways of ensuring high performance is creating a hardware/software combination. If hardware and software are created together and hardware is used to its fullest potential, entire system will work better. Also, error handling is important to make system perform well.

If there is an error and system continually tries to fix it with no actual solution, performance will suffer and time will be wasted. Sometimes it is just better to not try to fix an error and to just cut the losses. The setup of a new session should always be fast so terminating a session and starting a new one might be faster than trying to fix an error. Although loss could cost network owner, user should not be too upset with a quick termination and setup of a new session. There will always be some losses for network owner since no system is perfect but these losses should be kept as simple and tiny as possible. Hence, it is important to have a generic framework for a network because there are millions of devices that need to connect to LTE networks. There are many different network providers that device users can use. The network provider should not make a difference because the user can switch between them at any time but device should still work the same. Users pay for their usage of network. At any time, they should be able to send an SMS message, use Internet or make a call. All this usage must be tracked and charged correctly and efficiently. There are many different conditions to consider when charging a device's usage. This charging must happen constantly for all devices on the network, therefore performance is very important. Fast databases, powerful hardware, and efficient software must be used optimally together to achieve a system strong enough to handle hundreds of thousands of devices at any given time.

13.3.2   DIAMETER ROUTING AGENT

Diameter is an AAA protocol, it is an extensible control plane message exchange protocol designed as a successor protocol to RADIUS. With reference to OSI Layered model it comes in Application Layer. Diameter is a message-based protocol, where diameter signaling nodes exchange messages and receive acknowledgments for each message exchanged between the nodes.[14] At transport layer, it uses TCP/IP or SCTP (secured) as underlying transport protocol with TLS or DTLS as secure variants which makes diameter reliable. Diameter is successor of RADIUS, RADIUS is a AAA protocol based on UDP. TCP uses implicit hand-shaking mechanism for providing reliability, ordering, or data integrity, unlike UDP which does not have these features. Unreliability and small message size are the disadvantages of Radius. The session binding, load balancing, and traffic management is provided by Diameter Routing Agent (DRA) which is stated by 3GPP.[6,18] A DRA is also referred to as a Diameter Signaling Controller (DSC). At transport level, Diameter protocol uses either TCP/IP or SCTP. Two secure variations also exist: TLS for TCP/IP and DTLS for SCTP.

13.3.3 DIAMETER INTERFACES

It follows the peer-to-peer architecture and it has three different types of nodes called as clients, servers, and peers (or agents).

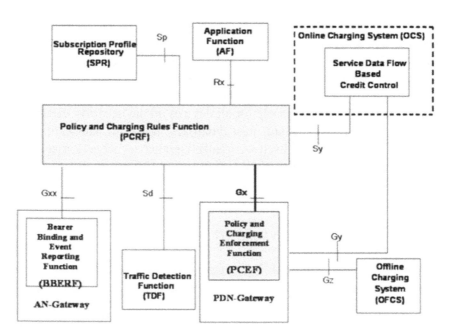

FIGURE 13.3 Diameter interfaces.[21]

Diameter NE (Network Element) that receives user connection request, (such as a network access server), is referred to as client. Diameter NE that processes request is referred to as server. Intermediary nodes are referred to as peers or agents as shown in Figure 13.3.

Diameter Clients—Diameter clients initiate requests by either connecting directly with servers or sending messages using intermediaries. It is a job of intermediate agent, a DRA, for example, to determine how to send message to destination. Typically, clients do not handle or process messages. Examples of a Diameter client include a network access server (NAS), packet gateway (P-GW), or MME.

Diameter Servers—Servers process messages generated by clients. There are many kinds of servers, including edge access systems, online charging systems, PCRF, emergency call routing services, and AAA systems.

Diameter Peers—Peers (or agents) are directly connected NEs between which messages are delivered. Not all messages are processed by a peer; a downstream intermediate agent, a DRA, may process the message.

13.3.4   *DevOps DELIVERY MODEL*

DevOps is the practice of operations, testers, and development people participating together in the entire software product lifecycle, from design to production support, with continuous feedback at each stage. The need for DevOps model came from increasing need for faster delivery of software product and need to innovate in delivery basis. DevOps is formulated to overcome shortcomings of agile system of software delivery model. DevOps model proposes to break the silos between development and operations. The idea of DevOps is to have a continuous feedback to improve software, this can be achieved by sharing responsibilities for developers, testers, and operations each other, traditional roles of these created silos between them. The automation of test execution cases and development of CI framework, this strategy helps in more interaction between development and quality teams in enhancing quality of software produced. Plans for inclusion of CI pipeline for Dynamic diameter engine (DDE) are discussed below. The following steps are to be followed for a successful CI implementation. Further continuous deployment and steps to be incorporated for moving to DevOps and CI are discussed in detail. The role of automation is important in implementation of CI.

CI pipeline for DDE.

- Each commit to Gerrit triggers a Jenkins job that starts the build. Code is allowed to commit only if the build passes.
- Nightly builds are triggered with a run of complete set of UT/MTs. If test fails, then mainline is blocked for check-ins, except for some "selective" commits that should be approved prior to check-in.

Continuous Deployment for DDE

- Every night, staging installation starts and application images are produced in a local lab.
- Target installation of application images is done and automated "Smoke" test cases are run.
- A mail is sent to complete DDE R&D team with test results and images are ready to be downloaded

Plan for Moving to DevOps and CI

- An "automation" framework "Rammbock" is being setup to run DDE functional test cases.
- Automation of functional test cases has started and the target for 100% automation.
- The complete "staging" and "deployment" setup should be moved to the vLabs provided by A&A DevOps team.
- The DDE software images will be built only after a successful run of the complete automated "functional" tests.
- The staging lab and target installation lab will be moved to vLAB.
- Target lab installation will be using the local scripts and not Arie's script; hence; there is no dependency on/dscmain for target installation scripts.

Currently, CI will be run on the vlab with heat install and report will be generated, once we start getting report and installation part is stable, we start using "heat stack create" for installation.

13.3.5 TEST AUTOMATION FRAMEWORK

Growth in emerging technologies means a requirement for quicker updates for product. To carry out releases for product in a much faster pace requires faster and better ways of testing the software for quality assurance. Test automation framework is a set of rules and standards followed by testers to increase code re-usability and reduce manual work, which are repetitive. Robot Framework is an open source test automation framework. Robot Framework has libraries written both in Java and Python. These libraries help in using some in-built keywords. Syntax for this framework is tabular and easy to use. Test cases are built using in-built keywords and further user-keywords can be built using in-built keywords.[8–10] Modular architecture of Robot Framework can be extended to create test libraries of our own. When suite execution gets started, the Robot Framework parses test data, and it then interacts with system/product under test by user-defined keywords. It can interact with system using other test tools as drivers. Tests can be executed from windows or command line, thus you can obtain log, output and report files in html as well as XML format. These log files help in debugging extensively and successful execution of each of the keywords in suite are also notified.

13.4   DESIGN AND SPECIFICATIONS OF AUTOMATION TOOL

There are two tools designed to enhance automation framework for the DDE software. First one is DDE Applications Counters.

13.4.1   COUNTERS

Counters are one of parameters used for Key Performance Indicator (KPI) calculations. KPIs are business metrics used by corporate executives to track and analysis various factors such as

- Network health.
- Performance of the product.

Verification of counter values at the end of test execution confirms success of test cases and desired performance of software.[13,16] There are numerous counter values defined according to the 3GPP standards. These counters are to be verified after each test case. Under current scenario of process continuous delivery of the DDE software, after automated execution of tests manual verification of counters using ganglia GUI for refresh interval of 15 min is done. Since data is huge and task of validating is cumbersome, need for designing an automation framework tool to validate these counters is required. As a solution to the above-mentioned problem, a tool is to be developed using ROBOT to verify these counters automatically. The following steps are to be done in process,

- Counters are generated in interval of every 15 min, logging the details of calls.
- Arguments are to be validated to ensure calls are executed properly.
- Hence, a counter validation tool is developed using RIDE to automate this task.
- During execution of test cases, the 3GPP counter files are fetched from the DDE at an interval of 15 min.
- Data from XML files are extracted and stored in a dictionary.
- Then values in dictionary are validated.

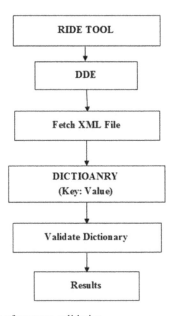

FIGURE 13.4   Flowchart of counter validation.

Figure 13.4 shows the flow diagram of execution of events in process of counter verification. Keyword for verification is called in test cases. A 3GPP XML file is generated in pilot node of DDE. These files contain updated counter values. These files need to be fetched and values are needed to be validated. Counter key-value pairs are stored in a dictionary, and further validated against user given key-value pairs. Robot Framework is used for writing user-defined keywords. The tool contains various user-defined keywords which are

- Get-Latest-3GPP-XML
- Getmultiplefiles
- GetElementsFrmXml-List
- Loop-ListToDict
- Validate-Counter

All these keywords mentioned here perform various operations required for validation of counters. Functionality of each of these keywords is explained below.

Get-Latest-3GPP-XML–The 3GPP files are logged in pilot node. Since various test cases are executed in a single test-suite, the latest 3GPP XML file should be fetched from pilot node. Get-Latest-3GPP-XML keyword performs this task. The following arguments are given as an input.

- PILOT_FLOAT_IP: The external IPv4 address of the DDE OAM Node
- Non-OAM-Type: APP /DB /IO (APP /DB /IO: VM Type)
- Num: Number of APP/DB/IO
- Counter_refreshing_time: Time interval for fetching of 3GPP files.

These above argument values are to be provided by user. IPv4 address of pilot is needed to remotely login into pilot machine from heat toolkit machine (these names are in reference with Openstack instances and nodes for computing and others). The XML file is fetched from pilot and transferred to heat machine (local machine) to store them into a dictionary.

Getmultiplefiles—The DDE consists of two Operations and maintenance (OAM) nodes, two Database (DB) nodes, and required number of Input output (IO) nodes, and Application nodes. XML files are generated in OAM node for all other nodes. Hence, files for all nodes should be fetched. This keyword iterates over number of nodes and calls another keyword to extract data from XML files. Following arguments can be given as an input,

- path: Path to XML file in local machine.
- num_nodes: To get total number of XML files.
- Idname: Virtual Network function (VNF) stack-ID of the node
- Non-OAM-Type: APP /DB /IO

This keyword calls GetElementsFrmXml-List to extract data from XML file.

GetElementsFrmXml-List—This keyword creates a list of key and value pairs and concatenate them to form a dictionary of key-value pairs, XML file contains multiple values for a single key, and hence latest should be updated. The following input arguments are used,

- path: Path to XML files in local machine.
- infolength: Number of arguments in the XML file.

It runs another keyword Loop-ListToDict to accumulate the key-values into dictionary.

Loop-ListToDict—This keyword loops over all the key-value pairs in the list of keys and their values. The dictionary is updated with values from list and values are updated, further different key-value pairs should be updated to the existing dictionary. The following are input arguments:

- meas_types_list: List contains arguments of counter.
- meas_results_list: List contains values of arguments.

This keyword updates the key-value pairs at each iteration.

13.5 IMPLEMENTATION

Implementation of this work involves three major stages. The initial step to begin with design and analysis, followed by implementation and testing. These steps are rigorously followed until design requirements are fulfilled. Various test scenarios are to be covered for a robust framework. Design parameters should be defined according to the requirements of framework. Attribute value pairs for diameter protocol in LTE network are also considered. Diameter peers are configured with service manager for testing framework. Dynamic routing agent (diameter routing agent) is brought up in VM using Openstack GUI. Basic calls are to be executed to test automation framework. The programming languages used are Python and Unix shell scripting. Robot-Framework-based keyword programming is used. Command line interface (CLI) is used to spawn diameter routing agent, service manager, diameter peers, realm, and to add rules for realm-based routing and host-based routing. Framework developed should be embedded into test-suite of CI-machine. Diameter peers should be configured, the automation framework developed is based on keywords, and required keywords are called in the test cases appropriately for validation and debugging after execution of test cases.

13.5.1 AUTOMATION FRAMEWORK

Robot Framework enables user to test on software with a keyword-based driven approach. Here end application is DDE. The ROBOT consists of the automation framework

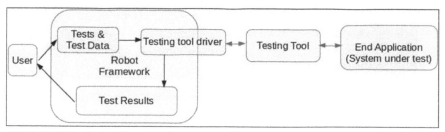

FIGURE 13.5   Robot Framework ecosystem.[22]

Figure 13.5 shows Robot Framework system workflow. Robot Framework interacts with system under test directly or by other means. Tests are written in RIDE terminal and execution of tests are conditional to code written. The code consists of a bunch of built-in and user-defined keywords. Test data is accessed by Robot Framework using resources. The test results are products of tests; they are used to determine results of tests. The logs and output files generated can be used to assess various portions of test. Test results contain logs and outputs of test case. These files give detailed report of execution of each individual user-defined keyword.

13.5.2   PROGRAMMING LANGUAGES

The programming languages used in this work are Python, Robot-keyword-based programming, Robot Framework is embedded by keywords, tools and libraries implemented using Python. Shell scripting is used to run commands on DDE nodes.

Hardware and software requirements—Figure 13.6 shows hardware specifications required for installation of DDE (Baremetal). Above fields are specified system memory, total number of cores, CPU specifications, and others.

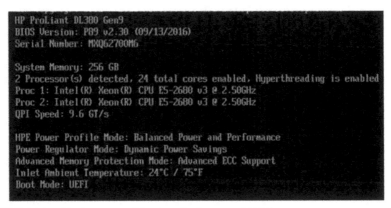

FIGURE 13.6   Hardware specifications for baremetal DDE.

Openstack—It is a open source, community driven, For public and private clouds a cloud management platform is used.[1,12] It is sometimes called as a cloud operating system (or cloud management platform (CMP)). DDE is implemented in cloud using Openstack nodes in NOVA (compute node) and networks in NEUTRON (network node). Inside the OpenStack many core projects are included one of them is OpenStack networking which is referred to as Neutron. Basically it is just the collection of application program interfaces and collection of many projects.

As shown in Figure 13.7, a web interface is provided which contains a dashboard that will give administrators control to their users to control. It also controls the big pool of storage, compute, and networking resources throughout the data center. It has many key projects such as NOVA, NEUTRON, SWIFT, GLANCE, KEYSTONE, CYNDER, and many others for different functioning.

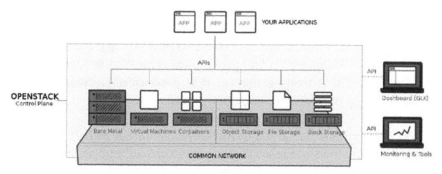

FIGURE 13.7   Overview of Open stack software.[23]

Robot Framework—It is an open source test automation framework. Robot Framework has libraries written in both Java and Python. These libraries help in using some in-built keywords. Syntax for this framework is tabular and easy to use. Test cases are built using in-built keywords and further user-keywords can be built using in-built keywords. The modular architecture of Robot Framework can be extended to create test libraries of our own. When suite execution gets started, Robot Framework parses test data, and it then interacts with system/product under test by the user-defined keywords. It can interact with system using other test tools as drivers.

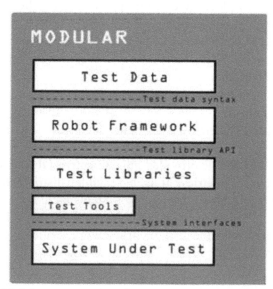

FIGURE 13.8   Modular representation of test automation using robot Framework.[24]

Robot Framework usage in software testing is shown in Figure 13.8. Robot Framework has many useful inbuilt libraries such as, Collections, Operating System, Process, String, Telnet, SSH, XML, and others.

Wireshark—Wireshark is a packet analyzer, it is open-source software thereof it is freely available, it is used for analyzing, network troubleshooting, development of communication protocol and software, education.

Wireshark allows user to see traffic which is present on that interface, it includes configured addresses of the interfaces and multicast/broadcast traffic as well as unicast traffic which is not sent to MAC address of network interface controller. Wireshark is better than tcpdump, because it has some integrated sorting, a graphical front-end and filtering options. Wireshark is a data capturing tool that can understand structure (encapsulation) of hundred different networking protocols.

Postman—It is a HTTP client testing powerful application, postman is used for testing web services which are based on JSON-REST. This Application maintains history API calls, so it can be easily loaded at later for response viewer. Postman includes API which makes into proper groups and they can be called as "collections." So the main aim of collection is to find and reuse of API requests. It also provides the history list in that response viewer can simply click the API and load it easily. There are many menus in many places like header presets, header fields, URL input fields are their throughout the app.

Win SCP—Windows Secure Copy (WinSCP) is an open-source secure copy protocol (SCP), FTP, WebDAV, and secure file transfer protocol (SFTP) client for Microsoft Windows. The function of Win SCP is to secure file transfer between a local and a remote computer machine. It provides the facility of synchronization of files and managing files. To provide the secure transfer, we use Secure Shell (SSH) and supports the SCP protocol in addition of SFTP.

CSF component—DDE provides an interactive GUI to view live or historical statistics of metrics. It is a scalable, distributed monitoring system for high-performance computing systems including clusters. DDE on bare-metal solution forms a ganglia cluster where both the bare-metal nodes are installed with the ganglia framework. Following are the use cases of Ganglia.

- Diameter Counters (list below) should be captured in Ganglia Web GUI 3gpp XML file generation for 5, 15, 30 min.
- System Counters should be captured in Ganglia Web GUI 3gpp XML file generation for System Counters.
- Default Collection Interval is set to 15 min.
- Ganglia GUI should open with URL with the respective IP address.

13.6 RESULTS AND DISCUSSION

After developing framework, it should be tested for validation. Different test cases are considered to validate results to authenticate working of automation framework. The test results are compared with those expected.

13.6.1 CLI DETAILS AND CONFIGURATION

CLI is an interface between machine and user to configure DDE. Various nodes in DDE are configured using Openstack commands for VNF. All the nodes are brought up using heat machine.

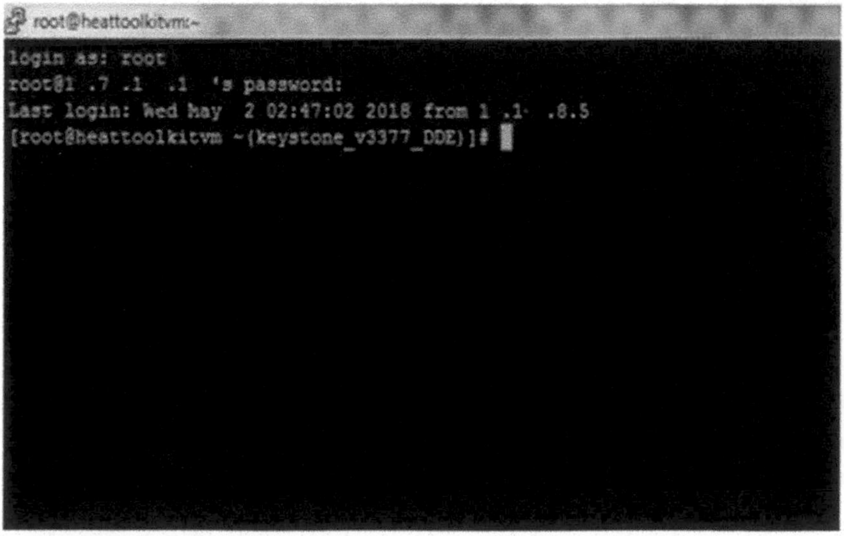

FIGURE 13.9   CLI mode login details in Linux machine.

Figure 13.9 shows the CLI mode login details in Heat machine (Linux operating system), to configure various DDE nodes and execute test cases and to do calls, heat machine is used. Heat machine can be accessed using SSH. The username and passwords for heat machine are configured using Service Manager Graphic User Interface.

13.6.2   TESTING THE TRACE TOOL

Tracing automation tool should be tested for its functionality and desired performance. Hence, for the following test cases trace tool is used and verified for expected results.

Test Case 1: To verify functionality of tracing tool for the test suite of Diameter-multiple outbound connections.

1	Enable-Tracing	10.75.148.113		TEST_YES	ddebvnf-ddeapp	TRACE
2	Start Diameter Server And Connect To Remote Client	ser1		server_origin_host=${server_orig	server_origin_realm=${destination	port=${server_origin_host1_port1
3	Start Diameter Server And Connect To Remote Client	ser2		server_origin_host=${server_orig	server_origin_realm=${destination	port=${server_origin_host1_port2
4	Start Client	clt1		client_origin_host=${client_origin_	client_origin_realm=${origin_realm	
5	Sleep	1				
6	Send SH					
7	Receive SH on any server and send SH answer					
8	Receive SH-Answer					
9	Disable-Tracing	10.75.148.113		ddebvnf-ddeapp	TRACE	

FIGURE 13.10   Snapshot of RIDE terminal using trace keywords.

Figure 13.10 shows the snapshot of trace automation tool being used in diameter multiple outbound connections. There are two keywords used in this scenario. Enable-tracing keyword starts tracing for that particular test case. Here Enable-tracing keywords take the pilot IP address (IPv4), node type, and the trace level. This keyword first searches for number of nodes present in DDE. Further iterates over each of those nodes specified.

Logging of files continues in background until the end of test case. The disable-tracing keyword triggers logged files to be sent back to heat machine. It kills all processes and further deletes logged files from pilot machine after files are transferred to heat machine as shown in Figure 13.11. So, that the memory is not drained. Figure 13.12 shows successful execution of keywords start logging.

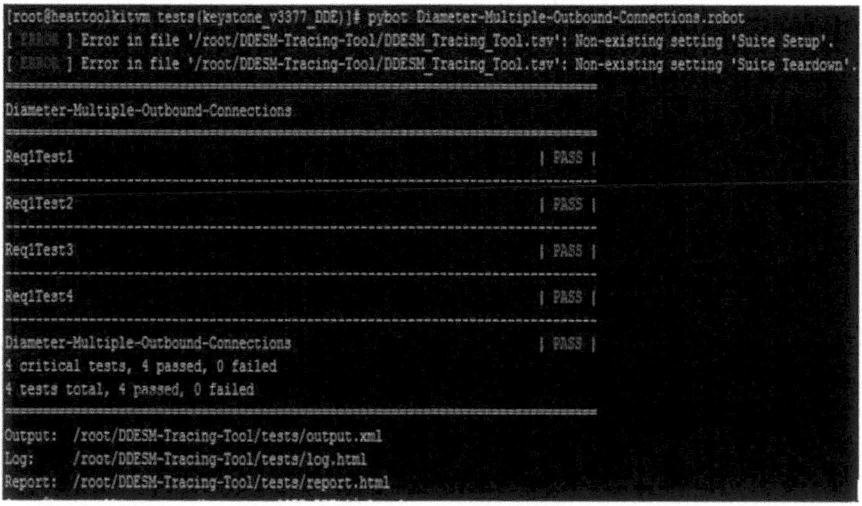

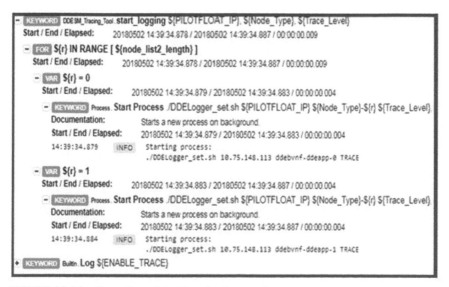

FIGURE 13.11   Execution of test cases in heat machine.

FIGURE 13.12   Execution of start logging keyword.

Keyword start process in figure executes a script DDELogger_setback. sh in background, which kills all tail and sets trace to default level of INFO, as shown in Figure 13.13. The start logging keyword starts a process in background, which needs to be terminated after the end of the execution of test case. The stop logging does this work; it kills all the processes started by start logging keyword. This ensures that pilot node is not overwhelmed with all the processes started, which in turn may result in failure of other test cases due to memory and other issues.

FIGURE 13.13   The execution of stop logging keyword.

FIGURE 13.14   Execution report of the test suite and its status.

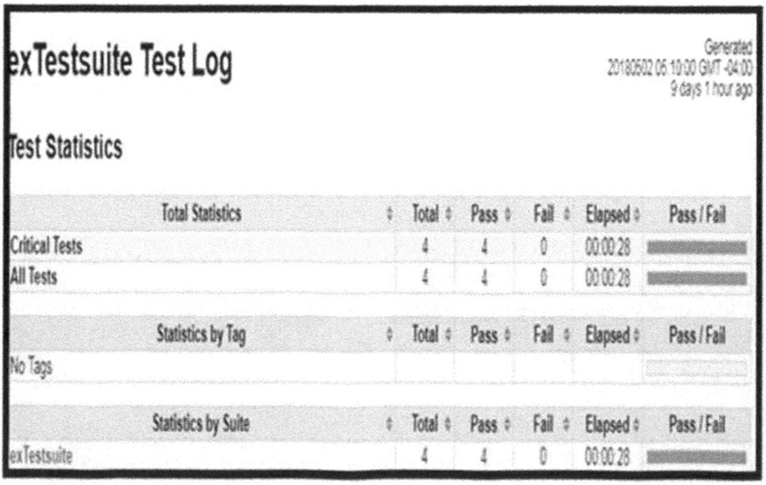

exTestsuite Test Log

Generated
2018/05/02 05:10:00 GMT -04:00
9 days 1 hour ago

Test Statistics

Total Statistics		Total	Pass	Fail	Elapsed	Pass / Fail
Critical Tests		4	4	0	00:00:28	▬▬▬▬▬
All Tests		4	4	0	00:00:28	▬▬▬▬▬
Statistics by Tag		**Total**	**Pass**	**Fail**	**Elapsed**	**Pass / Fail**
No Tags						
Statistics by Suite		**Total**	**Pass**	**Fail**	**Elapsed**	**Pass / Fail**
exTestsuite		4	4	0	00:00:28	▬▬▬▬▬

FIGURE 13.15   Test statistics for given test suite.

Figures 13.14 and 13.15 show test statistics for test suite executed. The test suite "exTestsuite" consists of four test cases. Tracing tool is enabled for all test statistics. The tracing tool stores log files in heat machine for all nodes present in DDE for given argument. The test statistics shows that all tests are executed successfully and time duration taken.

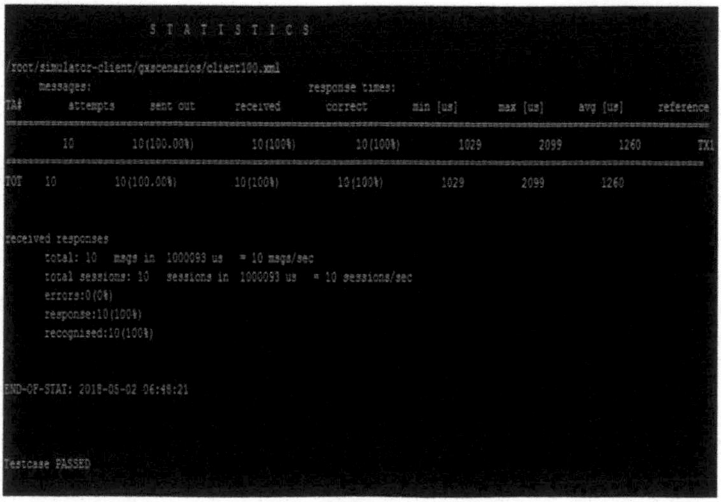

FIGURE 13.16   Statistics of call executed per single client.

Simulator is configured for 100 clients and to send 1000 messages per second. That makes the single client to send 10 messages per second. Figure 13.16 shows statistics from the simulator end. Following contents are mentioned after simulation, number of responses, total number of sessions, and number of errors.

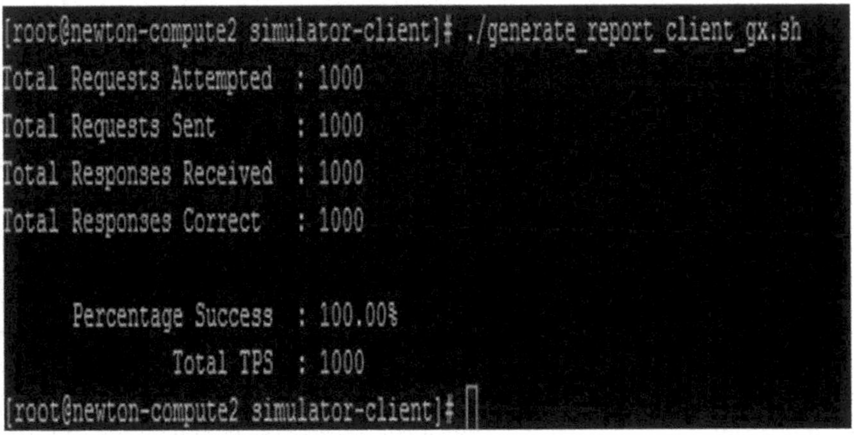

```
[root@newton-compute2 simulator-client]# ./generate_report_client_gx.sh
Total Requests Attempted  : 1000
Total Requests Sent       : 1000
Total Responses Received  : 1000
Total Responses Correct   : 1000


     Percentage Success  : 100.00%
          Total TPS  : 1000
[root@newton-compute2 simulator-client]#
```

FIGURE 13.17   Statistics of overall client responses sent from simulator.

Figure 13.17 shows the overall statistics from client end. Fields in figure shows numbers configured. The total number of requests attempted from clients is thousand, as configured. Moreover all thousand requests are sent to the DDE. Total responses received from DDE are shown here to be thousand and all the responses are deemed correct by client simulator. The total TPS is shown to be1000, which means 1000 messages are sent per second. All these values are to be validated against those obtained at the DDE.

Figure 13.18 shows output after the execution at RIDE (ROBOT IDE) terminal. Test execution shows test case is passed and the log output files are store in local directory. Dictionary is further logged at output terminal for verification.

Figure 13.19 shows dictionary key-value pairs being displayed at RIDE output terminal. The counter values are updated into dictionary to be used for validation against simulation results.

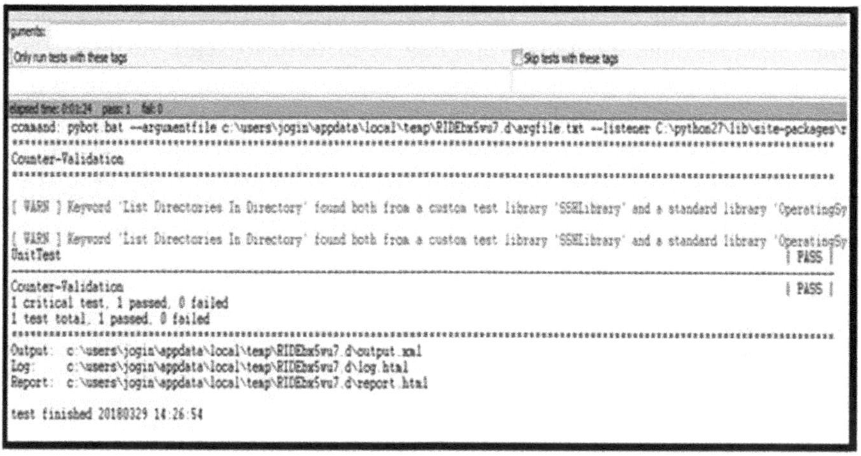

FIGURE 13.18   RIDE terminal output after the execution of test cases.

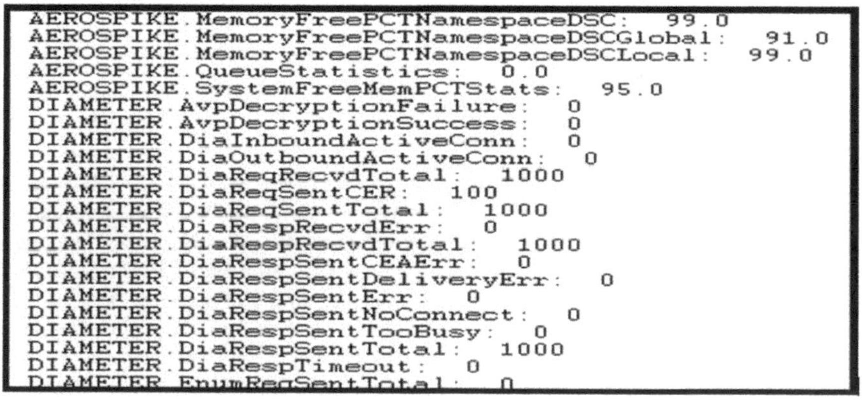

FIGURE 13.19   Output at RIDE terminal (dictionary key-value pairs).

13.7   CONCLUSION AND FUTURE DIRECTION

Automation framework of counters and tracing has resulted in an optimized testing. Framework is developed using ROBOT, which is OS independent, making it more flexible to adopt. The counter framework made task easier by providing instant output rather than looking graph/XML file manually. Tracing framework makes user to enable tracing quickly just by providing few inputs in less time which helps in debugging issue. The framework can be updated further according to requirements in future.

This framework has resulted in laying out standard automation framework further. The automation framework has been built to scale further. The new test cases can be automated and validated by current tools there by avoiding manual testing and wastage of time in doing repetitive tasks. The automation framework is embedded into CI pipeline, further this helps in the reduced software delivery and improved software quality, there by incorporating the DevOps model of software delivery.

KEYWORDS

- **automation**
- **DevOps**
- **dynamic diameter engine**
- **long term evolution**
- **3GPP**

REFERENCES

Ardagna, D. *et al.*, MODAClouds: A Model-Driven Approach for the Design and Execution of Applications on Multiple Clouds. In *2012 4th International Workshop on Modeling in Software Engineering (MISE)*, 2012; pp 50–56; DOI: 10.1109/MISE.2012.6226014.

Perera, P.; Silva, R.; Perera, I. Improve Software Quality through Practicing DevOps. In *2017 Seventeenth International Conference on Advances in ICT for Emerging Regions (ICTer)*, 2017; pp 1–6. DOI: 10.1109/ICTER.2017.8257807.

Ghantous, G. B.; Gill, A. The DevOps Reference Architecture Evaluation: A Design Science Research Case Study. In *2020 IEEE International Conference on Smart Internet of Things (SmartIoT)*, 2020; pp 295–299. DOI: 10.1109/SmartIoT49966.2020.00052.

Baral, K.; Mohod, R.; Flamm, J.; Goldrich, S.; Ammann, P. Evaluating a Test Automation Decision Support Tool. In *2019 IEEE International Conference on Software Testing, Verification and Validation Workshops (ICSTW)*, 2019; 69–76. DOI 10.1109/ICSTW.2019.00034.

Bencomo, N.; Belaggoun, A.; Issarny, V. Dynamic Decision Networks for Decision-Making in Self-Adaptive Systems: A Case Study. In *2013 8th International Symposium on Software Engineering for Adaptive and Self-Managing Systems (SEAMS)*, 2013; pp 113–122. DOI: 10.1109/SEAMS.2013.6595498.

3rd Generation Partnership Project, Technical Specification Group Services and System Aspects, Policy and Charging Control (PCC) Architecture, *3GPP TS 23.203, Release 10*, Sept 2010. http://www.3gpp.org

Bang, S. K.; Chung, S.; Choh, Y.; Dupuis, M. A Grounded Theory Analysis of Modern Web Applications: Knowledge, Skills, and Abilities for DevOps. In *Proceedings of the 2nd Annual Conference on Research in Information Technology (RIIT '13)*; Association for Computing Machinery: New York, NY, USA, 2013; pp 61–62. DOI: https://doi. org/10.1145/2512209.2512229.

Achieving the Full Potential of Test Automation. *White Paper, USA—2015 Pioneer Ct.* http://www.logigear.com

Data Driven Testing with Keywords. White Paper, *USA—2015 Pioneer Ct.* http://www. logigear.com

Offshore Software Test Automation, White Paper, *USA—2015 Pioneer Ct.* http://www. logigear.com

Lorenz, G.; Moore, T.; Manes, G.; Hale, J.; Shenoi, S. Securing SS7 Telecommunications Networks. In *Proceedings of the 2001 IEEE Workshop on Information Assurance and Security United States Military Academy*; West Point: NY, 5–6, June 2001; pp 273–278.

Hosono, S. A DevOps Framework to Shorten Delivery Time for Cloud Applications. *Int. J. Comput. Sci. Eng.* **2012,** 329–344. DOI: https://doi.org/10.1504/IJCSE.2012.049753

Fitzpatrick, L.; Dillon, M The Business Case for DevOps: A Five-Year Retrospective. In *Cutter IT Journal* 2011, *24* (8), 19–27.

Diameter Signalling Control Annual Worldwide and Regional Market Size and Forecasts. https://www.marketsandmarkets.com/PressReleases/diameter-signaling.asp

IRMC: 6201.2 SEC Course Notes, National Défense University, Information Resources Management College, Washington, DC. https://www.ndu.edu/

Test Data Management in Software Testing Life Cycle—Business Need and Benefits in Functional, Performance, and Automation Testing. White paper. https://www.infosys. com/services/it-services/white-papers/documents/test-data-management-software.pdf

Kottapalli, N. Diameter and LTE Evolved Packet System. *Radisys, White Paper.*

Pallavi, G. S. *et al.*, Cap to Diameter protocol converter. In *2015 International Conference on Green Computing and Internet of Things (ICGCIoT)*, 2015; pp 598–602. DOI: 10.1109/ICGCIoT.2015.7380535.

https://www.3glteinfo.com/lte-resources-free-system-posters/

https://blog.3g4g.co.uk/2012/01/lte-policy-control-and-charging.html

http://www.lteandbeyond.com/2012/10/the-sy-interface-between-pcrf-and-ocs.html

https://subscription.packtpub.com/book/application_development/9781783283033/1/ ch01lvl1sec10/the-robot-framework-ecosystem

https://www.includehelp.com/cloud-computing/openstack-the-future-of-next-generation-private-cloud.aspx

https://www.edureka.co/blog/robot-framework-tutorial/

Index